The
SCIENTIFIC AMERICAN

Book of
The Brain

The Editors of
Scientific American

Introduction by Antonio R. Damasio

The Lyons Press
Guilford, Connecticut
An imprint of The Globe Pequot Press

To buy books in quantity for corporate use
or incentives, call **(800) 962–0973**
or e-mail **premiums@GlobePequot.com.**

The Lyons Press is an imprint of The Globe Pequot Press.

Designed by Compset, Inc.

ISBN 978-1-58574-285-1

The Library of Congress Cataloging-in-Publication Data

The Scientific American book of the brain / from the editors of
  Scientific American ; introduction by Antonio Damasio.
        p.   cm.
   ISBN 1-5874-285-6 (pb)   ISBN 1-55821-965-X (hc)
   1. Cognitive neuroscience.   2. Brain.   3. Neuropsychology.
4. Neuropsychiatry.   5. Brain-Diseases. I. Damasio, Antonio R.
II. Scientific American.   III. Title: Book of the brain.
QP360.5.S35 2001
612.8'2-dc21   99-39387
                                                    CIP

Manufactured in the United States of America

First edition/ Sixth printing

The

SCIENTIFIC AMERICAN

Book of

The Brain

CONTENTS

INTRODUCTION

The Scientific American *Book of the Brain* is the scientific equivalent of a rich collection of short stories by several great writers—the sort of collection that you can open to any page and find immediate reading pleasure. I am not suggesting, of course, that you begin reading this book in the middle or in reverse, but I am saying that you can read it any way you wish, from the beginning chapter to the last or section by section or in random order. All the chapters touch on some intriguing aspect of neuroscience, all the chapters describe in accessible language an interesting problem and its state-of-the-art solution, and each chapter is written by scientists, clinicians, or thinkers who are among the leading experts on the respective topic. The very fact that it is possible to assemble this volume is a tribute to *Scientific American* and to its commitment to offering readers an intelligible account of current science prepared by those who practice the science itself.

One reason why this volume is so valuable has to do with its theme: the infinitely fascinating human brain, and, in most chapters, one aspect or another of the infinitely fascinating human mind. There is nothing new about the brain as object of fascination, and the deep connection between brain and mind has long been a source of puzzlement. The enterprise we now call neuroscience has long been an attractive and respected field of inquiry. And yet something entirely unexpected has been happening in recent years, and that something is responsible for the current excitement in the sciences of the brain and mind. Over a brief period of time, roughly twenty years, there has been an explosive development of new theories regarding the brain and its workings, as well as powerful techniques that allow us to study the brain experimentally—from the level of the individual nerve cells and the molecules those cells require in order to operate to the level of the brain's macrosystems.

The combination of theoretical and technological advances has permitted an astounding accumulation of facts regarding the structure and function of the nervous system, approached from many angles and levels of its organization. And although the main drive behind this massive inquiry was the simple and perfectly justified desire to understand the mysterious brain, the effort ended up permitting a scientifically based assault on the diseases of the brain. The Decade of the Brain, during which, it has been said, we have learned more than during the century that preceded it, is also the decade during which it became possible to daydream about the prevention of Alzheimer's disease and Parkinson's disease, and about the radical management of learning disorders such as dyslexia.

The chapters collected in this volume cover the waterfront of contemporary neuroscience and its applied medical fields: neurology and psychiatry. You can read about techniques such as magnetic resonance scanning—known as MR or MRI—which allows you to see the brain's macro-anatomy in ways previously possible only at the time of autopsy; or positron emission tomography—known as PET—which lets you glimpse the activity of a small chunk of brain caught in the process of contributing to some grand mental function. You can read about the techniques of molecular neurobiology that allow you to discover how genes assemble neural circuits and how genes are engaged by the activity of circuits and help govern processes such as learning. As a result of the availability of those techniques, you can also read about a new understanding of what lies beneath some of the brain's fundamental products: emotion, memory, language. The detail that is now available in the understanding of the neural basis of these operations would have been unthinkable just a decade ago.

You can read about controversial topics that may have intrigued you in the daily news: Are the brains of men and women equal or different? Are there identifiable brain correlates for homosexuality? What are false memories and how are they created? What does the IQ test really measure? How does this measure relate to creativity? How does intelligence in general, whatever intelligence is, relate to the genes responsible for assembling a given brain?

And you can read about the recent medical applications of so much progress in neuroscience: new treatments for Parkinson's disease, new drugs to manage depression and mania, the brave new world of rationally designed drugs that will control so many diseases in the future.

At the simultaneous finish of the millennium, the century, and the Decade of the Brain, this volume lets the reader take stock of where neuroscience stands, for a fleeting moment, before the start of the next thousand years. The volume closes, quite appropriately, with three chapters on the matter of consciousness—the very last topic to be added to the agenda of neuroscience and one about which, not surprisingly, little consensus has developed so far. The next Scientific American *Book of the Brain* will probably begin with three chapters on consciousness and will summarize the agreements that will have been reached over the phenomena of consciousness. In fact, if another ten years go by, the entire book may be entirely about the biology of consciousness.

—Antonio R. Damasio

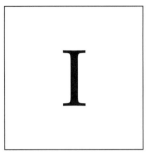

Mapping the Brain

THE DEVELOPING BRAIN

During fetal development, the foundations of the mind
are laid as billions of neurons form appropriate connections
and patterns. Neural activity and stimulation are crucial
in completing this process

Carla J. Shatz

An adult human brain has more than 100 billion neurons. They are specifically and intricately connected with one another in ways that make possible memory, vision, learning, thought, consciousness, and other properties of the mind. One of the most remarkable features of the adult nervous system is the precision of this wiring. No aspect of the complicated structure, it would appear, has been left to chance. The achievement of such complexity is even more astounding when one considers that during the first few weeks after fertilization many of the sense organs are not even connected to the embryonic processing centers of the brain. During fetal development, neurons must be generated in the right quantity and location. The axons that propagate from them must select the correct pathway to their target and finally make the right connection.

How do such precise neural links form? One idea holds that the brain wires itself as the fetus develops, in a manner analogous to the way a computer is manufactured: that is, the chips and components are assembled and connected according to a preset circuit diagram. According to this analogy, a flip of a biological switch at some point in prenatal life turns on the computer. This notion would imply that the brain's entire structure is recorded in a set of biological blueprints—presumably DNA—and that the organ begins to work only after the wiring is essentially complete.

Research during the past decade shows that the biology of brain development follows very different rules. The neural connections elaborate themselves from an immature pattern of wiring that only grossly approximates the adult pattern. Although humans are born with almost all the neurons they will ever have, the mass of the brain at birth is only about one fourth that of the adult brain. The brain becomes bigger because neurons grow in size, and the number of axons and dendrites as well as the extent of their connections increases.

Workers who have studied the development of the brain have found that to achieve the precision of the adult pattern, neural function is necessary: the brain must be stimulated in some fashion. Indeed, several observations during the past few decades have shown that babies who spent most of their first year of life lying in their cribs developed abnormally slowly. Some of these infants could not sit up at 21 months of age, and fewer than 15 percent could walk by about the age of three. Children must be stimulated—through touch, speech, and images—to develop fully. Based in part on such observations, some people favor enriched environments for young children, in the hopes of enhancing development. Yet current studies provide no clear evidence that such extra stimulation is helpful.

Much research remains to be done before anyone can conclusively determine the types of sensory input that encourage the formation of particular neural connections in newborns. As a first step toward understanding the process, neurobiologists have focused on the development of the visual system in other animals, especially during the neonatal stages. It is easy under the conditions that prevail at that stage to control visual experience and observe behavioral response to small changes. Furthermore, the mammalian eye differs little from species to species. Another physiological fact makes the visual system a productive object of study: its neurons are essentially the same as neurons in other parts of the brain. For these reasons, the results of such studies are very likely to be applicable to the human nervous system as well.

But perhaps the most important advantage is that in the visual system, investigators can accurately correlate function with structure and identify the pathway from external stimulus to physiological response. The response begins when the rods and cones of the retina transform light into neural signals. These cells send the signals to

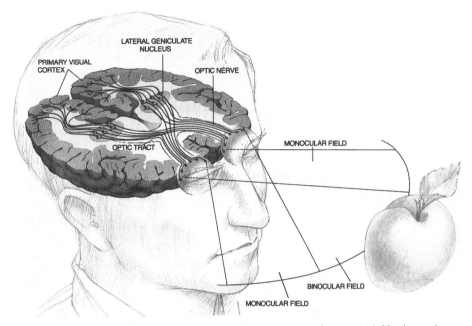

Visual pathway in the adult demonstrates the segregation of axons. Neighboring retinal ganglion cells in each eye send their axons to neighboring neurons in the lateral geniculate nucleus. Similarly, the neurons of the geniculate nucleus map their axons onto the visual cortex. The system forms a topographically orderly pattern that in part accounts for such characteristics as binocular vision. (Dana Burns-Pizer)

the retinal interneurons, which relay them to the output neurons of the retina, called the retinal ganglion cells. The axons of the retinal ganglion cells (which make up the optic nerve) connect to a relay structure within the brain known as the lateral geniculate nucleus. The cells of the lateral geniculate nucleus then send the visual information to specific neurons located in what is called layer 4 of the (six-layer) primary visual cortex. This cortical region occupies the occipital lobe in each cerebral hemisphere [*see illustration above*].

Within the lateral geniculate nucleus, retinal ganglion cell axons from each eye are strictly segregated: the axons of one eye alternate with those from the other and thus form a series of eye-specific layers. The axons from the lateral geniculate nucleus in turn terminate in restricted patches within cortical layer 4. The patches corresponding to each eye interdigitate with one another to form structures termed ocular dominance columns.

To establish such a network during development, axons must grow long distances, because the target structures form in different regions. The retinal ganglion cells are generated within the eye. The lateral geniculate neurons take shape in an embryonic structure known as the diencephalon, which will form the thalamus and hypothalamus. The layer 4 cells are created in another protoorgan called the telen-cephalon, which later develops into the cerebral cortex. From the beginning of fe-tal development, these three structures are many cell-body diameters distant from one another. Yet after identifying one or the other of these targets, the axons reach it and array themselves in the correct topographic fashion—that is, cells located near one another in one structure map their axons to the correct neighboring cells within the target.

This developmental process can be compared with the problem of stringing tele-phone lines between particular homes located within specific cities. For instance, to string wires between Boston and New York, one must bypass several cities, in-cluding Providence, Hartford, New Haven, and Stamford. Once in New York, the lines must be directed to the correct borough (target) and then to the correct street address (topographic location).

Corey Goodman of the University of California at Berkeley and Thomas Jessel of Columbia University have demonstrated that in most instances, axons immediately recognize and grow along the correct pathway and select the correct target in a highly precise manner. A kind of "molecular sensing" is thought to guide growing axons. The axons have specialized tips, called growth cones, that can recognize the proper pathways. They do so by sensing a variety of specific molecules laid out on the surface of, or even released from, cells located along the pathway. The target it-self may also release the necessary molecular cues. Removing these cues (by genetic or surgical manipulation) can cause the axons to grow aimlessly. But once axons have arrived at their targets, they still need to select the correct address. Unlike pathway and target selection, address selection is not direct. In fact, it involves the correction of many initial errors.

The first hint that address selection is not precise came from experiments using radioactive tracers. Injections of these tracers at successively later times in fetal de-velopment outline the course and pattern of axonal projections. Such studies have also shown that structures emerge at different times in development, which can further complicate address selection.

For instance, Pasko Rakic of Yale University has shown that in the visual pathway in monkeys, the connections between the retina and the lateral geniculate nucleus appear first, followed by those between the lateral geniculate nucleus and layer 4 of the visual cortex. Other studies found that in cats and primates (including hu-mans), the lateral geniculate nucleus layers develop during the prenatal period, be-fore the rods and cones of the retina have formed (and thus before vision is even possible). When Simon LeVay, Michael P. Stryker, and I were postdoctoral fellows at Harvard Medical School, we found that at birth, layer 4 columns in cats do not even exist in the visual cortex. I subsequently determined that even earlier, in fetal life, the cat has no layers in the lateral geniculate nucleus. These important visual structures emerge only gradually and at separate stages.

The functional properties of neurons, like their structural architecture, do not attain their specificity until later in life. Microelectrode recordings from the visual cortex of newborn cats and monkeys reveal that the majority of layer 4 neurons respond equally well to visual stimulation of either eye. In the adult, each neuron in layer 4 responds primarily if not exclusively to stimulation of one eye only. This finding implies that during the process of address selection, the axons must correct their early "mistakes" by removing the inputs from the "inappropriate" eye.

In 1983 my colleague Peter A. Kirkwood and I found further evidence that axons must fine-tune their connections. It came from our work on the brains of six-week-old cat fetuses (the gestation period of the cat is about nine weeks). We removed a significant portion of the visual pathway—from the ganglion cells in both eyes to the lateral geniculate nucleus—and placed it in vitro in a special life-support chamber. (Inserting microelectrodes in a fetus is extremely difficult.) The device kept the cells alive for about 24 hours. Next we applied electrical pulses to the two optic nerves to stimulate the ganglion cell axons and make them fire action potentials, or nerve signals. We found that neurons in the lateral geniculate nucleus responded to the ganglion cells and, indeed, received inputs from both eyes. In the adult the layers respond only to stimulation of the appropriate eye.

The eventual emergence of discretely functioning neural domains (such as the layers and ocular dominance columns) indicates that axons do manage to correct their mistakes during address selection. The selection process itself depends on the branching pattern of individual axons. In 1986 David W. Sretavan, then a doctoral student in my laboratory, was able to examine the process in some detail. Experimenting with fetal cats, he selectively labeled single retinal ganglion cell axons in their entirety—from the cell body in the retina to their tips within the lateral geniculate nucleus—at successively later stages.

He found that at the earliest times in development, when ganglion cell axons have just arrived within the lateral geniculate nucleus (after about five weeks of gestation), the axons assume a very simple sticklike shape and are tipped with a growth cone. A few days later the axons arising from both eyes acquire a "hairy" appearance: they have short side branches along their entire length.

The presence of side branches at this age implies that the inputs from both eyes mix with one another. In other words, the neural regions have yet to take on the adult structure, in which each eye has its own specific regions. As development continues, the axons sprout elaborate terminal branches and lose their side branches. Soon individual axons from each eye have highly branched terminals that are restricted to the appropriate layer. Axons from one eye that traverse territory belonging to those from the other eye are smooth and unbranched [*see illustration below*].

The sequence of developmental changes in the branching patterns shows that the adult pattern of connections emerges as axons remodel by the selective withdrawal and growth of different branches. Axons apparently grow to many different addressees within their target structures and then somehow eliminate addressing errors.

One possible explanation for axonal remodeling is that specific molecular cues are arrayed on the surface of the target cells. Although this idea might seem conceptually attractive, it has very little experimental support. An alternative explanation appears to be stronger. It holds that all target neurons are fair game. Then, some kind of competition between inputs would lead to formation of specific functional areas.

An important clue concerning the nature of the competitive interactions between axons for target neurons has come from the experiments of David H. Hubel

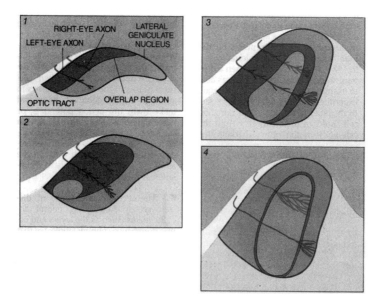

Axonal remodeling in the lateral geniculate nucleus occurs largely before birth. At the earliest times in development (1), the axons from the left eye and right eye are simple and tipped with growth cones. The shaded region represents the intermixing of inputs from both eyes. After further development (2), the axons grow many side branches. The axons soon begin to lose some side branches and start to extend elaborate terminal branches (3). Eventually these branches occupy the appropriate territory to form eye-specific layers (4). (Dana Burns-Pizer)

of Harvard Medical School and Torsten N. Wiesel of the Rockefeller University. In the 1970s, when both workers were at Harvard, they studied the formation of childhood cataracts. Clinical observations indicated that if the condition is not treated promptly, it can lead to permanent blindness in the obstructed eye. To emulate the effect, Hubel and Wiesel closed the eyelids of newborn cats. They discovered that even a week of sightlessness can alter the formation of ocular dominance columns. The axons from the lateral geniculate nucleus representing the closed eye occupy smaller than normal patches within layer 4 of the cortex. The axons of the open eye occupy larger than normal patches.

The workers also showed that the effects are restricted to a critical period. Cataracts, when they occur in adulthood and are subsequently corrected by surgery, do not cause lasting blindness. Apparently the critical period has ended long ago, and so the brain's wiring cannot be affected.

These observations suggest that the ocular dominance columns form as a consequence of use. The axons of the lateral geniculate nucleus from each eye somehow compete for common territory in layer 4. When use is equal, the columns in the two eyes are identical; unequal use leads to unequal allotment of territory claimed in layer 4.

How is use translated into these lasting anatomic consequences? In the visual system, use consists of the action potentials generated each time a visual stimulus is converted into a neural signal and is carried by the ganglion cell axons into the brain. Perhaps the effects of eye closure on the development of ocular dominance columns occur because there are fewer action potentials coming from the closed eye. If that is the case, blockage of all action potentials during the critical period of postnatal life should prevent axons from both eyes from fashioning the correct patterns and lead to abnormal development in the visual cortex. Stryker and William Harris, then a postdoctoral fellow at Harvard, obtained this result when they used the drug tetrodotoxin to block retinal ganglion cell action potentials. They found that the ocular dominance columns in layer 4 failed to appear (the layers in the lateral geniculate nucleus were unaffected because they had already formed in utero).

Nevertheless, action potentials by themselves are not sufficient to create the segregated patterns in the cortex. Neural activity cannot be random. Instead it must be defined, both temporally and spatially, and must occur in the presence of special kinds of synapses. Stryker and his associate Sheri Strickland, who are both at the University of California at San Francisco, have shown that simultaneous, artificial stimulation of all the axons in the optic nerves can prevent the segregation of axons from the lateral geniculate nucleus into ocular dominance columns within layer 4. Although this result resembles that achieved with tetrodotoxin, an important difference exists. Here ganglion cell action potentials are present—but all at the same

DEVELOPMENT AND NEURAL FUNCTION

One of the characteristics of the developing visual system is segregation of inputs: each eye adopts its own territory in the visual cortex. The process, however, can be completed only if the neurons are stimulated. In experiments with cat eyes, for example, the axons of the left eye and of the right eye overlap in layer 4 of the visual cortex at birth (*a*). Visual stimuli will cause the axons to separate and form ocular dominance columns in the cortex (*b*). Such normal development can be blocked with injections of tetrodotoxin; as a result, the axons never segregate, and the ocular dominance columns fail to emerge (*c*). Another way to perturb development is to keep one eye closed, depriving it of stimulation. The axons of the open eye then take over more than their share of territory in the cortex (*d*).

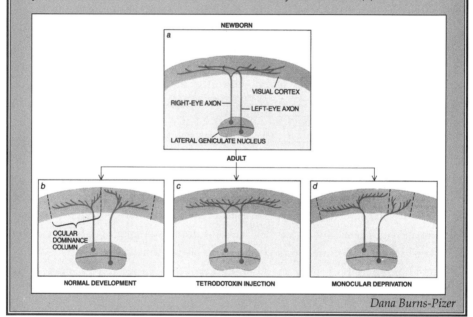

Dana Burns-Pizer

time. Segregation to form the columns in the visual cortex, on the other hand, proceeds when the two nerves are stimulated asynchronously.

In a sense, then, cells that fire together wire together. The timing of action-potential activity is critical in determining which synaptic connections are strengthened and retained and which are weakened and eliminated. Under normal circumstances, vision itself acts to correlate the activity of neighboring retinal ganglion cells, because the cells receive inputs from the same parts of the visual world.

What is the synaptic mechanism that strengthens or weakens the connections? As long ago as 1949, Donald O. Hebb of McGill University proposed the existence of special synapses that could execute the task. The signal strength in such synapses would increase whenever activities in a presynaptic cell (the cell supplying the synaptic input) and in a postsynaptic cell (the cell receiving the input) coincide. Clear evidence showing that such "Hebb synapses" exist comes from studies of the phenomenon of long-term potentiation in the hippocampus. Researchers found that the pairing of presynaptic and postsynaptic activity in the hippocampus can cause incremental increases in the strength of synaptic transmission between the paired cells. The strengthened state can last from hours to days.

Such synapses are now thought to be essential in memory and learning. Studies by Wolf J. Singer and his colleagues at the Max Planck Institute for Brain Research in Frankfurt and by Yves Fregnac and his colleagues at the University of Paris also suggest that Hebb synapses are present in the visual cortex during the critical period, although their properties are not well understood.

Just how coincident activity causes long-lasting changes in transmission is not known. There is general agreement among researchers that the postsynaptic cell must somehow detect the coincidence in the incoming presynaptic activity and in turn send a signal back to all concurrently active presynaptic inputs. But this cannot be the whole story. During the formation of the ocular dominance columns, inputs that are not active at the same time are weakened and eliminated.

Consequently, one must also propose the existence of a mechanism for activity-dependent synaptic weakening. This weakening—a kind of long-term depression—would occur when presynaptic action potentials do not accompany postsynaptic activity. Synapses that have this special property (opposite to that of Hebb synapses) have been found in the hippocampus and cerebellum. The results of the Stryker and Strickland experiments suggest that such synapses are very likely to exist in the visual cortex as well.

A strongly similar process of axonal remodeling operates as motor neurons in the spinal cord connect with their target muscles. In the adult, each muscle fiber receives input from only one motor neuron. But after motor neurons make the first contacts with the muscle fibers, each muscle fiber receives inputs from many motor neurons. Then, just as in the visual system, some inputs are eliminated, giving rise to the adult pattern of connectivity. Studies have shown that the process of elimination requires specific temporal patterns of action-potential activity generated by the motor neurons.

The requirement for specific spatial and temporal patterns of neuronal activity might be likened to a process whereby telephone calls are placed from addresses in one city (the lateral geniculate nucleus in the visual system) to those in the next city (the visual cortex) to verify that connections have been made at the correct locations. When two near neighbors in the lateral geniculate nucleus simultaneously call neighboring addresses in the cortex, the telephones in both those homes will ring. The concurrent ringing verifies that relations between neighbors have been preserved during the wiring process.

If, however, one of the neighbors in the lateral geniculate nucleus mistakenly makes connections with very distant parts of layer 4 or with parts that receive input from the other eye, the called telephone will rarely if ever ring simultaneously with those of its neighbors. This dissonance would lead to the weakening and ultimate removal of that connection.

The research cited thus far has explored the remodeling of connections after the animal can move or see. But what about earlier in development? Can mechanisms of axonal remodeling operate even before the brain can respond to stimulation from the external world? My colleagues and I thought the formation of layers in the lateral geniculate nucleus in the cat might be a good place to address this question. After all, during the relevant developmental period, rods and cones have not yet emerged. Can the layers develop their specific territories for each eye even though vision cannot yet generate action-potential activity?

We reasoned that if activity is necessary at these early times, it must somehow be generated spontaneously within the retina, perhaps by the ganglion cells themselves. If so, the firing of retinal ganglion cells might be contributing to layer construction, because all the synaptic machinery necessary for competition is present. It should be possible to prevent the formation of the eye-specific layers by blocking action-potential activity from the eyes to the lateral geniculate nucleus.

To hinder activity during fetal development, Sretavan and I, in collaboration with Stryker, implanted special minipumps containing tetrodotoxin in utero just before the lateral geniculate nucleus layers normally begin to form in the cat (at about six weeks of fetal development). After two weeks of infusion, we assessed the effects on the formation of layers. Much to our satisfaction, the results of these in utero infusion experiments showed clearly that the eye-specific layers do not appear in the presence of tetrodotoxin. Moreover, by examining the branching patterns of individual ganglion cell axons after the treatment, we reassured ourselves that tetrodotoxin did not simply stunt normal growth.

In fact, the branching patterns of these axons were very striking. Unlike normal axons at the comparable age, the tetrodotoxin-treated axons did not have highly restricted terminal branches. Rather they had many branches along the entire length of the axon. It was as if, without action-potential activity, the information necessary to withdraw side branches and elaborate the terminal branches was missing.

In 1988, at about the same time these experiments were completed, Lucia Galli-Resta and Lamberto Maffei of the University of Pisa achieved the extraordinary technical feat of actually recording signals from fetal ganglion cells in utero. They demonstrated directly that retinal ganglion cells can indeed spontaneously generate bursts of action potentials in the darkness of the developing eye. This observation, taken together with our experiment, strongly suggests that action-potential activity is not only present but also necessary for the ganglion cell axons from the two eyes to segregate and form the eye-specific layers.

Still, there must be constraints on the spatial and temporal patterning of ganglion cell activity. If the cells fired randomly, the mechanisms of correlation-based, activity-dependent sorting could not operate. Furthermore, neighboring ganglion cells in each eye somehow ought to fire in near synchrony with one another, and the firing of cells in the two eyes, taken together, should be asynchronous. In addition, the synapses between retinal ganglion cell axons and neurons of the lateral geniculate nucleus should resemble Hebb synapses in their function: they should be able to detect correlations in the firing of axons and strengthen accordingly.

We realized that to search for such patterns of spontaneous firing, it would be necessary to monitor simultaneously the action-potential activity of many ganglion cells in the developing retina. In addition, the observation had to take place as the eye-specific layers were developing. A major technical advance permitted us to achieve this goal. In 1988 Jerome Pine and his colleagues at the California Institute of Technology, among them doctoral student Markus Meister, invented a special multielectrode recording device. It consisted of 61 recording electrodes arranged as a flat, hexagonal array. Each electrode can detect action potentials generated in one to several cells. When Meister arrived at Stanford University to continue post-doctoral work with Denis Baylor, we began a collaboration to see whether the electrode array could be used to detect the spontaneous firing of fetal retinal ganglion cells.

In these experiments, it was necessary to remove the entire retina from the fetal eye and place it, ganglion-cell-side down, on the array. (It is technically impossible to put the electrode array itself into the eye in utero.) Rachel Wong, a postdoctoral fellow from Australia visiting my laboratory, succeeded in carefully dissecting the retinas and in tailoring special fluids necessary to maintain the living tissue for hours in a healthy condition.

When neonatal ferret retinas were placed on the multielectrode array, we simultaneously recorded the spontaneously generated action potentials of as many as 100 cells. The work confirmed the in vivo results of Galli-Resta and Maffei. All cells on the array fired within about five seconds of one another, in a predictable and rhythmic pattern. The bursts of action potentials lasted several seconds and were followed by long silent pauses that persisted from 30 seconds to two minutes. This observation showed that the activity of ganglion cells is indeed correlated. Further

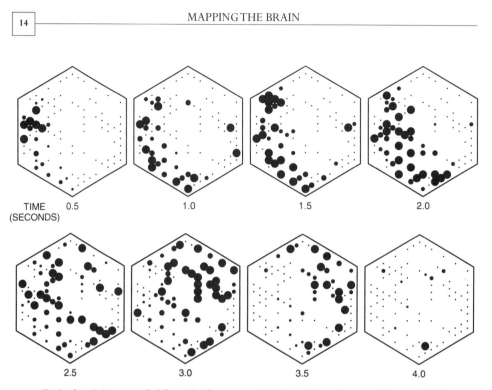

TIME (SECONDS)   0.5   1.0   1.5   2.0

2.5   3.0   3.5   4.0

Retinal activity, recorded frame by frame every 0.5 second by a hexagonal array of microelectrodes (small dots), is locally synchronized. Each diagram represents the pattern and intensity of action-potential firing (large dots) of individual ganglion cells. The wave of retinal activity sweeps across from the lower left to the top right of the retina. (Guilbert Gates/JSD)

analysis demonstrated that the activity of neighboring cells is more highly correlated than that of distant cells on the array.

Even more remarkable, the spatial pattern of firing resembled a wave of activity that swept across the retina at about 100 microns per second (about one tenth to one hundredth the speed of an ordinary action potential) [*see illustration above*]. After the silent period, another wave was generated but in a completely different and random direction. We found that these spontaneously generated retinal waves are present throughout the period when eye-specific layers take shape. They disappear just before the onset of visual function.

From an engineering standpoint, these waves seem beautifully designed to provide the required correlations in the firing of neighboring ganglion cells. They also ensure a sufficient time delay, so that the synchronized firing of ganglion cells remains local and does not occur across the entire retina. Such a pattern of firing could help refine the topographic map conveyed by ganglion cell axons to each eye-specific layer. Moreover, the fact that wave direction appears to be entirely random implies that ganglion cells in the two eyes are highly unlikely ever to fire synchronously—a requirement for the formation of the layers.

Future experiments will disrupt the waves in order to determine whether they are truly involved in the development of connections. In addition, it will be important to determine whether the correlations in the firing of neighboring ganglion cells can be detected and used by the cells in the lateral geniculate nucleus to strengthen appropriate synapses and weaken inappropriate ones. This seems likely, since Richard D. Mooney, a postdoctoral fellow in my laboratory, in collaboration with Daniel Madison of Stanford, has shown that long-term potentiation of synaptic transmission between retinal ganglion cell axons and the lateral geniculate nucleus neurons is present during these early periods of development. Thus, at present, we can conclude that even before the onset of function, ganglion cells can spontaneously fire in the correct pattern to fashion the necessary connections.

Is the retina a special case, or might many regions of the nervous system generate their own endogenous activity patterns early in development? Preliminary studies by Michael O'Donovan of the National Institutes of Health suggest that the activity of motor neurons in the spinal cord may also be highly correlated very early in development. It would appear that activity-dependent sorting in this system as well might use spontaneously generated signals. Like those in the visual system, the signals would refine the initially diffuse connections within targets.

The necessity for neuronal activity to complete the development of the brain has distinct advantages. The first is that, within limits, the maturing nervous system can be modified and fine-tuned by experience itself, thereby providing a certain degree of adaptability. In higher vertebrates, this process of refinement can occupy a protracted period. It can begin in utero and, as in the primate visual system, continue well into neonatal life, where it plays an important role in coordinating inputs from the two eyes. The coordination is necessary for binocular vision and stereoscopic depth perception.

Neural activity confers another advantage in development. It is genetically conservative. The alternative—exactly specifying each neural connection using molecular markers—would require an extraordinary number of genes, given the thousands of connections that must be formed in the brain. Using the rules of activity-dependent remodeling described here is far more economical. A major challenge for the future will be to elucidate the cellular and molecular bases for such rules.

—September 1992

The Visual Image
in Mind and Brain

*In analyzing the distinct attributes of images, the brain
invents a visual world. Unusual forms of blindness show what
happens when specialized parts of the cortex malfunction*

Semir Zeki

The study of the visual system is a profoundly philosophical enterprise; it entails an inquiry into how the brain acquires knowledge of the external world, which is no simple matter. The visual stimuli available to the brain do not offer a stable code of information. The wavelengths of light reflected from surfaces change along with alterations in the illumination, yet the brain is able to assign a constant color to them. The retinal image produced by the hand of a gesticulating speaker is never the same from moment to moment, yet the brain must consistently categorize it as a hand. An object's image varies with distance, yet the brain can ascertain its true size.

The brain's task, then, is to extract the constant, invariant features of objects from the perpetually changing flood of information it receives from them. Interpretation is an inextricable part of sensation. To obtain its knowledge of what is visible, the brain cannot therefore merely analyze the images presented to the retina; it must actively construct a visual world. To do so, the brain has developed an elaborate neural mechanism, one so marvelously efficient that it took a century of study before anyone even guessed at its many components. Indeed, when studies of cerebral diseases began to reveal some secrets of the visual brain, neurologists initially dismissed the startling implications as improbable.

The hallmark of that machinery is a complex division of labor. It is manifested anatomically in discrete cortical areas and subregions of areas specialized for particular visual functions; it is manifested pathologically in an inability to acquire knowledge about some aspect of the visual world when the relevant machinery is specifically compromised. Paradoxically, none of this subdivision and specialization within the brain is normally evident at the perceptual level. The visual cortex thus presents us with the intellectual challenge of trying to understand how its components cooperate to give us a unified picture of the world, one that bears no trace of the division of labor within it. There is, to use an old phrase, a great deal more to vision than meets the eye.

This modern conception of the visual brain has evolved only within the past two decades. The early neurologists, starting with those who worked during the late 19th century, saw it very differently. Laboring under the false notion that objects transmitted visual codes in reflected or emitted light, they thought that an image was "impressed" on the retina, much as it would be on a photographic plate. These retinal impressions were subsequently transmitted to the visual cortex, which served to analyze the contained codes. This decoding process led to "seeing." Understanding what was seen—making sense of the received impressions and resolving them into visual objects—was thought to be a separate process that arose through the association of the received impressions with similar ones experienced previously.

This view of how the brain operates, which persisted into the mid-1970s, was therefore also deeply philosophical, although neurologists never acknowledged it as such. It divided sensing from understanding and gave each faculty a separate seat in the cortex. The origin of this dualistic doctrine is obscure, but it bears a resemblance to Immanuel Kant's belief in the two faculties of sensing and of understanding, the former passive and the latter active.

Neurologists saw evidence for their supposition in the fact that the retina connects overwhelmingly to one distinct part of the brain, the striate or primary visual cortex, also known as area V1. This connection is made with high topographic precision: V1 effectively contains a map of the entire retinal field. The retina and V1 are linked through a subcortical structure called the lateral geniculate nucleus, which contains six layers of cells. The four uppermost layers con-

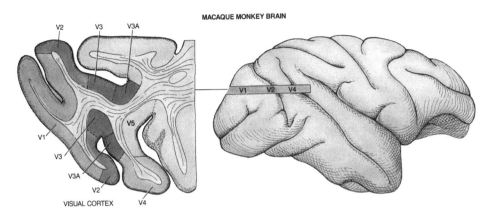

Visual cortex of the macaque monkey has been studied in detail. A cross section through the brain (left) at the level indicated (right) shows part of the primary visual cortex (V1) and some of the other visual areas in the prestriate cortex (V2–V5). (Carol Donner)

tain cells with small cell bodies and are referred to as the parvocellular layers; the two lowest layers, which have large cell bodies, are the magnocellular layers. Many years ago the late neurologist Salomon E. Henschen of Uppsala University supposed that the function of the large cells was "collecting light" and that of the small cells was registering colors; his basic insight, that the anatomic subdivisions had functional implications, has assumed increasing importance in recent years.

Neurologists at the time found that lesions anywhere along the pathway connecting the retina with V1 created a field of absolute blindness, the extent and position of which corresponded precisely with the size and location of the lesion in V1. That observation led Henschen to conceive of area V1 as the "cortical retina"—the place where "seeing" occurred.

Moreover, the German psychiatrist Paul Emil Flechsig of Leipzig University had shown during the late 19th century that certain regions of the brain, among them V1, had a mature appearance at birth, whereas others, including the cortical regions surrounding V1, continued to develop, as though their maturation depended on the acquisition of experience. For Flechsig and most other neurologists, this observation implied that V1 was "the entering place of the visual radiation into the organ of the psyche," and the areas around it were the repositories of higher "psychic" functions *(Cogitationzentren)* related to sight. Flechsig's theory found support in rather questionable evidence purporting to show that lesions in this so-called visual association cortex, unlike those in V1, might lead to "mind blindness" *(Seelenblind-*

heit), a condition in which subjects were thought to see but not to comprehend what they saw.

Surprisingly, it was research on the visual association cortex that ultimately compromised this dualistic concept of visual brain organization. Work undertaken in the 1970s by John M. Allman and Jon H. Kaas of the University of Wisconsin in the owl monkey and by me in the macaque monkey showed that the visual association cortex—now better referred to as the prestriate cortex—consists of many different cortical areas separated from V1 by another area, V2. A turning point in our understanding of how the brain constructs the visual image came subsequently, with my demonstration that these areas are individually specialized to undertake different tasks [*see illustration on previous page*].

In my physiological studies, I presented macaques with a range of stimuli (colors, lines of various orientations, and dots moving in different directions) and, using electrodes, monitored the activity of cells in the prestriate cortex. The results showed that all the cells in a prestriate area called V5 are responsive to motion, that most are directionally selective and that none is concerned with the color of the moving stimulus. These facts suggested to me that V5 is specialized for visual motion. (Neuroanatomic terminology is not always uniform; some investigators prefer the label MT to V5.)

In contrast, I found that the overwhelming majority of cells in another area, V4, are to some extent selective for specific wavelengths of light and that many are selective for line orientation, the constituents of form, as well. By far most of the cells in two further adjoining areas, V3 and V3A, are also selective for form but like the cells of area V5 are largely indifferent to the color of the stimulus.

These studies led me to propose in the early 1970s the concept of functional specialization in the visual cortex, which supposes that color, form, motion, and possibly other attributes of the visible world are processed separately. Because the preponderance of input to the specialized areas comes from V1, a corollary of this finding was that V1 must also show a functional specialization, as must area V2, which receives input from V1 and connects with the same specialized areas. These two areas must, in a sense, act as a kind of post office, parceling out different signals to the appropriate areas.

In recent years, new tissue-staining techniques in combination with physiological studies have provided a startling confirmation of that theory. They have also allowed us to trace these specializations from V1 throughout the prestriate cortex.

With the advent of positron emission tomography (PET), which can measure increases in regional cerebral blood flow when people perform specific tasks, my colleagues at the Hammersmith Hospital in London and I have begun to apply these findings, which were derived from experiments on monkeys, to a direct study of the human brain. We found that when normal-seeing humans view a Land color Mondrian (an abstract painting containing no recognizable objects), the highest increase in regional cerebral blood flow occurs in a structure named the fusiform gyrus. By analogy with a similar region in macaque monkeys, we refer to this cortical area as human V4. The results are very different when subjects view a pattern of moving black-and-white squares: the highest cerebral blood flow then occurs in a more lateral area, quite separate from V4, which we call human V5.

This demonstration of the separation of motion and color processing constitutes direct evidence that functional specialization is also a feature of the human visual cortex. The PET studies reveal another interesting feature: under both conditions of stimulation, area V1 (and probably the adjoining area V2) also showed marked increases in regional cerebral blood flow. As in the monkeys, these regions, too, must be distributing signals to different areas of the prestriate cortex.

The key to the distribution system in these areas lies in their structural and functional organization. Area V1 is unusually rich in cell layers, yet it reveals an even richer architecture if one examines it with a staining technique first applied by Margaret Wong-Riley of the Medical College of Wisconsin in Milwaukee. The organelles known as mitochondria contain a metabolic enzyme called cytochrome oxidase that makes energy available to a cell. By staining a region of the brain for that enzyme, researchers can identify which cells have the greatest metabolic activity.

When so stained, the metabolic architecture of V1 is characterized by columns of cells that extend from the cortical surface to the underlying nerve tissue called white matter. If viewed in sections cut parallel to the cortical surface, these columns appear as heavily stained blobs or puffs, separated from one another by more lightly stained interblob regions. At Harvard Medical School, Margaret Livingstone and David H. Hubel found that wavelength-selective cells are concentrated in the blobs of V1, whereas form-selective cells are concentrated in the interblobs.

The columns are especially prominent in the second and third layers of V1, which receive input from the parvocellular layers of the lateral geniculate nucleus. The cells in those parts of the lateral geniculate nucleus respond in a strong, sustained way to visual stimuli, and many of them are concerned with color.

A distinct set of structures can be seen in layer 4B of V1, which receives input from the magnocellular layers of the lateral geniculate nucleus, whose cells respond transiently to stimuli and are mostly indifferent to color. Layer 4B projects to areas V5 and V3. The cells in layer 4B that connect with V5 are clustered into small patches that are isolated by cells connected to other visual areas. In short, the organization of layer 4B in V1 suggests that certain parts of it are specialized for motion perception and are segregated from regions that handle other attributes.

Like V1, area V2 has a special metabolic architecture. In the case of V2, however, that architecture takes the shape of thick stripes and thin stripes separated from one another by more lightly staining interstripes. As work done by Edgar A. DeYoe and David C. Van Essen of the California Institute of Technology, by Hubel and Livingstone, and by Stewart Shipp of University College, London, and me has shown, cells selective for wavelength congregate in the thin stripes, and cells selective for directional motion are found in the thick stripes. Cells sensitive to form are distributed in both the thick stripes and the interstripes.

V1 and V2 might therefore be said to contain pigeonholes into which the different signals are assembled before being relayed to the specialized visual areas. The cells in these pigeonholes have small receptive fields; that is, they respond only to stimuli falling on a small region of the retina. They also register information about only a specific attribute of the world within that receptive field. It is as though V1 and V2 were undertaking a piecemeal analysis of the entire field of view.

These facts allow us to delineate four parallel systems concerned with different attributes of vision—one for motion, one for color, and two for form. The two that are computationally most distinct from each other are the motion and color systems. For the motion system, the pivotal prestriate area is V5; its inputs run from the retina, through the magnocellular layers of the lateral geniculate nucleus, to layer 4B of V1. From there the signals pass to V5, both directly and through the thick stripes of V2. The color system depends on area V4; its inputs pass through the parvocellular layers of the lateral geniculate nucleus to the blobs of V1, then proceed to V4 directly or through the thin stripes of V2. [*See color plate 1*].

Of the two form systems, one is intimately linked to color, and the other is independent of it. The first is based on V4 and derives its inputs from the parvocellular layers of the lateral geniculate nucleus by way of the interblobs of V1 and the interstripes of V2. The second is based on V3 and is more concerned with dynamic form—the shapes of objects in motion. It derives its inputs from the magnocellular layers of the lateral geniculate nucleus through layer 4B of V1; the signals then proceed to V3 both directly and through the thick stripes of V2.

Although these four systems are distinct, the anatomy of areas V1 and V2 offers many opportunities for the pigeonholes to communicate with one another, as do the direct connections between the specialized visual areas. Hence, there is an admixture of the parvocellular and magnocellular signals, which the prestriate areas use in different ways to execute their functions.

This remarkable segregation of functions is reflected in some of the pathologies afflicting the visual cortex. Lesions in specific cortical areas produce correspondingly specific visual syndromes that may be far less debilitating than total blindness yet still severe enough to drive patients to distraction and despair. Lesions in area V4 lead to achromatopsia, in which patients see only in shades of gray. This syndrome is different from simple color blindness: not only do such patients fail to see or know the world in color, they cannot even recall colors from a time before the lesion formed. Nevertheless, if their retinas and V1 regions are healthy, their knowledge of form, depth, and motion remains intact [*see color plate 1*].

Similarly, a lesion in area V5 produces akinetopsia, in which patients neither see nor understand the world in motion. While at rest, objects may be perfectly visible to them, but motion relative to them causes the objects to vanish. The other attributes of vision remain unscathed—a specificity that results directly from functional differentiation in the human visual cortex.

Given the separation of form and color in the cortex, it is perhaps a little surprising that no one has ever reported a complete and specific loss of form vision. A partial explanation is that such a deficit would require the obliteration of areas V3 and V4 to eliminate both form systems. Area V3 forms a ring around V1 and V2. Consequently, a lesion large enough to destroy all of V3 and V4 would almost certainly destroy V1 as well and thus cause total blindness.

Some patients with lesions in the prestriate cortex do suffer from a degree of form imperception, often coupled with achromatopsia. These people commonly experience far greater difficulty when identifying stationary forms than when the same forms are in motion. They frequently prefer watching television to watching the real world, because television is dominated by moving images. When faced with stationary objects, these patients often resort to the strategy of moving their heads to simplify the task of identification. These observations suggest that they acquire their knowledge of forms through the dynamic form system based in area V3.

The functional specialization in the visual cortex also manifests itself in a syndrome that I have called the chromatopsia ("color vision") of carbon monoxide poisoning. This condition has been described sporadically but not infrequently in the medical literature, although it was never taken seriously until the functional specialization was discovered. Some people who survive the lethal effects of smoke inhalation during fires often suffer diffuse cortical damage from carbon monoxide poisoning, which deprives tissues of oxygen. As a result, these patients often have vision that is severely compromised in all respects except one: their color vision is affected only mildly if at all. Because color is the only kind of visual knowledge available to them, the patients try—often unsuccessfully—to identify all objects solely on the basis of color. They may, for example, misidentify all blue objects as "ocean."

The precise cause of this strange chromatopsia is unknown. The metabolically active blobs of V1 and the thin stripes of V2, both of which are concerned with color, do have unusually high concentrations of blood vessels nourishing them. It is therefore probable that these regions are relatively spared from damage because their rich blood supply renders them less vulnerable to oxygen deprivation.

In summary, then, we know that a total lesion in area V1 produces a complete inability to acquire any visual information and that a lesion in one of the specialized areas makes a corresponding attribute of the visual world inaccessible and incomprehensible. What would happen, we might ask, if signals from the lateral geniculate nucleus were routed directly to the specialized areas, thus bypassing V1 altogether? Nature has actually done that experiment for us, and the resulting phenomenon provides more important insights into the functioning of the visual cortex.

The phenomenon is known as blindsight; it was first described by Ernst Pöppel of the University of Munich and his colleagues and later studied in great detail by Lawrence Weiskrantz of the University of Oxford and his colleagues. People with the condition are totally blind because of lesions in area V1. Yet if they are forced to guess, they can discriminate correctly among a wide variety of visual stimuli. They can, for example, distinguish between motion in different directions or between different wavelengths of light. Their abilities are imperfect and not completely reliable, but they are better than random guessing. Nevertheless, blindsight patients are not consciously aware of having seen anything at all, and they are often surprised that their "guesses" should have been so accurate.

The basis for this discrimination almost certainly resides in a small but direct connection between the lateral geniculate nucleus and the prestriate cortex, as uncovered by Masao Yukie of the Tokyo Metropolitan Institute for Neurosciences and by Wolfgang Fries of the University of Munich. Alternatively, some other as yet undiscovered subcortical connection to the specialized areas may be responsible. In any event, neurologists have good reason to suppose that in blindsight patients, visual signals reach the prestriate cortex.

Blindsight patients are people who "see" but do not "understand." Because they are unaware of what they have seen, they have not acquired any knowledge. In short, their "vision," which can be elicited only in laboratory situations, is quite useless. Thus, for the visual cortex to do its job of acquiring a knowledge of the world, a healthy V1 area is essential. V1 (and, by extension, V2) may be necessary because it begins to process information for further refinement by the specialized areas or because the results of the processing performed by the specialized areas are referred back to it.

The clinical literature holds many other examples that illuminate how the preprocessing in areas V1 and V2 may contribute directly and explicitly to perception.

Damage to V5 can destroy the ability to discriminate the direction or coherence of motions. Yet as Robert F. Hess of the University of Cambridge and his colleagues found, such akinetopsic patients may still be aware that motion of some type is taking place, presumably because of signaling by cells in V1 and V2 (and possibly in other areas that receive magnocellular pathway signals). Similarly, an achromatopsic patient with a V4 lesion, whom I studied with Fries, could discriminate between different wavelengths because of his largely intact V1 even though he could no longer interpret the wavelength information as color.

Further insights come from a comparison of the residual form-vision capacities in two other patients who have cortical lesions that are, in a way, complementary. The first patient has a diffuse cortical lesion, caused by carbon monoxide poisoning, that affects area V1. He has terrific difficulty copying even simple forms, such as geometric shapes or letters of the alphabet, because the form-detecting system in his V1 area is so severely compromised.

The second patient has an extensive prestriate lesion from a stroke that has generally spared area V1. He can reproduce a sketch of St. Paul's Cathedral with greater skill than many normal people, although it takes him a great deal of time to do so. Yet this patient has no comprehension of what he has drawn. Because his V1 system is largely intact, he can identify the local elements of form, such as angle and simple shapes, and accurately copy the lines he sees and understands. The prestriate lesion, however, prevents him from integrating the lines into a complex whole and recognizing it as a building. The patient sees and understands only what the limited capacity of his intact system allows.

The residual capacity in such patients unmasks an important feature of the organization of the visual cortex, namely, that none of the visual areas—not even "post office" areas V1 and V2—serves merely to relay signals to other areas. Instead each is part of the machinery that actively transforms the incoming signals and may contribute explicitly, if incompletely, to perception.

The profound division of labor within the visual cortex naturally raises the question of how the specialized areas interact to provide a unified image. The simplest way would be for all the specialized areas to communicate the results of their operations to one master area, which would then synthesize the incoming information. Philosophically, that solution begs the question, because one must then ask who or what looks at the composite image, and how it does so. That problem is beside the point, however, because the anatomic evidence shows no single master area to which all the antecedent areas exclusively connect. Instead the specialized areas connect with one another, either directly or through other areas.

For example, areas V4 and V5 connect directly and reciprocally with each other. Both of them also project to the parietal and temporal regions of the brain, but as my work with my colleagues has shown, the outputs from each area occupy their own unique territory within the receiving region. Direct overlap between the signals from V4 and V5 is minimal. It is as if the cortex wishes to maintain the separation of the distinct visual signals—a strategy it also employs in memory and other systems. Any integration of the signals within the parietal or temporal regions must occur through local "wiring" that connects the inputs.

In fact, integration of the visual information is a monumental task that necessitates a vast network of anatomic links between the four parallel systems at every level, because each level contributes explicitly to perception. Integration also creates some formidable problems. To understand coherent motion, for example, the brain must determine which features in the field of view are moving in the same direction and at the same speed. The motion-sensing cells in the specialized areas are able to make those comparisons because they have larger receptive fields than do their antecedent counterparts in V1.

Yet because their receptive fields are larger, these cells are inherently less efficient at pinpointing the position of any one stimulus within the visual field. For the brain to make spatial sense of the integrated information, the information must somehow be referred to an area that has a more precise topographic map of the retina and hence of the visual field. Of all the visual areas, the one with the most precise map is area V1, followed by V2. The specialized areas must therefore send information back to V1 and V2 so that the results of the comparisons can be mapped back onto the visual field.

Reentrant connections, which allow information to flow both ways between different areas, are also essential for resolving conflicts between cells that have different capabilities and are responding to the same stimulus. A good example of such conflicts is found in the responses of cells in V1 and V2 to illusory contours, such as those of the Kanizsa triangle.

In this famous illusion, a normal observer perceives a triangle among the presented shapes even though the lines forming the triangle are incomplete; the brain creates lines where there are none. As Rudiger von der Heydt and Esther Peterhans of the Zurich University Hospital have demonstrated, the form-selective cells in V1 do not respond to the illusion and do not signal that a line is present. The V2 cells receive their inputs from V1, but because the V2 cells have larger receptive fields and more analytic functions, they do respond to the illusion by "inferring" the presence of a line. To settle the conflict, the V2 cells must have reentrant inputs to their counterparts in V1.

Another difficulty that arises from the process of integration is the binding problem. Cells responding to the same object in the field of view may be scattered throughout V1. Something must therefore bind together the signals from those

Kaniza triangle stimulates neurons that code explicitly for such illusory contours.

cells so that they are treated as belonging to the same object and not separate ones. The problem becomes even more thorny when cells in two or more visual areas respond to different attributes of the same object.

One way to resolve the problem is for the cells to fire in temporal synchrony. In practice, such synchrony does occur, at least to some extent, among cells that are anatomically connected to one another, as work by Wolf J. Singer of the Max Planck Institute of Brain Research in Frankfurt and his colleagues has shown. We must then face the problem, however, of who or what determines that the firing should be synchronous. Reentrant inputs provide at least a partial solution by linking the output of one area to other areas that sent it information.

These problems have led me and my colleagues to develop a theory of multistage integration. It hypothesizes that integration does not occur in a single step through a convergence of output onto a master area, nor does integration have to be postponed until all the visual areas have completed their individual operations. Instead the integration of visual information is a process in which perception and comprehension of the visual world occur simultaneously.

The anatomic requirement for multistage integration is immense, because it involves reentrant connections between all the specialized areas as well as with areas V1 and V2, which feed them signals. Our studies indicate that such a network of reentrant connections does exist.

The reentrant inputs to areas V1 and V2 differ fundamentally from their forward connections to the specialized areas. The forward projections are patchy and discrete, because segregated groups of cells in V1 and V2 send their outputs to specialized areas with corresponding visual attributes. The return projections, however, are diffuse and fairly nonspecific. For example, whereas V5 receives input only from select groups of cells in layer 4B of V1, the return input from V5 to layer 4B is diffuse and encompasses the territory of all the cells in the layer, including ones that project into V3. This reentrant system can thus serve three purposes simultaneously: it can unite and synchronize the signals for form and motion found in two different visual pathways; it can refer information about motion back to an area with an accurate topographic map; it can integrate motion information from V5 with form information on its way to V3.

Similarly, whereas the output from V2 to the specialized areas is highly segregated, the return input from those areas to V2 is diffuse. V4 projects back not only to the thin stripes and the interstripes, from which it receives its input, but also to the thick stripes, from which it does not. This reentrant system can therefore help unite signals dealing with form, motion, and color.

It is becoming increasingly evident that the entire network of connections within the visual cortex, including the reentrant connections to V1 and V2, must function healthily for the brain to gain complete knowledge of the external world. Yet as patients with blindsight have shown, knowledge cannot be acquired without consciousness, which seems to be a crucial feature of a properly functioning visual apparatus. Consequently, no one will be able to understand the visual brain in any profound sense without tackling the problem of consciousness as well.

The past two decades have brought neurologists many marvelous discoveries about the visual brain. Moreover, they have led to a powerful conceptual change in our view of what the visual brain does and how it accomplishes its functions. It is no longer possible to divide the process of seeing from that of understanding, as neurologists once imagined, nor is it possible to separate the acquisition of visual knowledge from consciousness. Indeed, consciousness is a property of the complex neural apparatus that the brain has developed to acquire knowledge.

Thus, our inquiry into the visual brain takes us into the very heart of humanity's inquiry into its own nature. This is not to say that understanding the workings of the visual brain will resolve the problem of consciousness—far from it. But it is a good beginning.

—September 1992

BRAIN AND LANGUAGE

A large set of neural structures serves to represent concepts;
a smaller set forms words and sentences.
Between the two lies a crucial layer of mediation

Antonio R. Damasio and Hanna Damasio

What do neuroscientists talk about when they talk about language? We talk, it seems, about the ability to use words (or signs, if our language is one of the sign languages of the deaf) and to combine them in sentences so that concepts in our minds can be transmitted to other people. We also consider the converse: how we apprehend words spoken by others and turn them into concepts in our own minds.

Language arose and persisted because it serves as a supremely efficient means of communication, especially for abstract concepts. Try to explain the rise and fall of the communist republics without using a single word. But language also performs what Patricia S. Churchland of the University of California at San Diego aptly calls "cognitive compression." It helps to categorize the world and to reduce the complexity of conceptual structures to a manageable scale.

The word "screwdriver," for example, stands for many representations of such an instrument, including visual descriptions of its operation and purpose, specific instances of its use, the feel of the tool, or the hand movement that pertains to it. Or there is the immense variety of conceptual representations denoted by a word such as "democracy." The cognitive economies of language—its facility for pulling together many concepts under one symbol—make it possible for people to establish ever more complex concepts and use them to think at levels that would otherwise be impossible.

In the beginning, however, there were no words. Language seems to have appeared in evolution only after humans and species before them had become adept at generating and categorizing actions and at creating and categorizing mental representations of objects, events, and relations. Similarly, infants' brains are busy representing and evoking concepts and generating myriad actions long before they utter their first well-selected word and even longer before they form sentences and truly use language. However, the maturation of language processes may not always depend on the maturation of conceptual processes, since some children with defective conceptual systems have nonetheless acquired grammar. The neural machinery necessary for some syntactic operations seems capable of developing autonomously.

Language exists both as an artifact in the external world—a collection of symbols in admissible combinations—and as the embodiment in the brain of those symbols and the principles that determine their combinations. The brain uses the same machinery to represent language that it uses to represent any other entity. As neuroscientists come to understand the neural basis for the brain's representations of external objects, events, and their relations, they will simultaneously gain insight into the brain's representation of language and into the mechanisms that connect the two.

We believe the brain processes language by means of three interacting sets of structures. First, a large collection of neural systems in both the right and left cerebral hemispheres represents nonlanguage interactions between the body and its environment, as mediated by varied sensory and motor systems—that is to say, anything that a person does, perceives, thinks, or feels while acting in the world.

The brain not only categorizes these nonlanguage representations (along lines such as shape, color, sequence, or emotional state), it also creates another level of representation for the results of its classification. In this way, people organize objects, events, and relationships. Successive layers of categories and symbolic representations form the basis for abstraction and metaphor.

Second, a smaller number of neural systems, generally located in the left cerebral hemisphere, represent phonemes, phoneme combinations, and syntactic rules for combining words. When stimulated from within the brain, these systems assemble word-forms and generate sentences to be spoken or written. When stimulated externally by speech or text, they perform the initial processing of auditory or visual language signals.

COMPONENTS OF A SOUND-BASED LANGUAGE	
Phonemes	The individual sound units, whose concatenation, in a particular order, produces morphemes.
Morphemes	The smallest meaningful units of a word, whose combination creates a word. (In sign languages the equivalent of a morpheme is a visuomotor sign.)
Syntax	The admissible combinations of words in phrases and sentences (called grammar, in popular usage).
Lexicon	The collection of all words in a given language. Each lexical entry includes all information with morphological or syntactic ramifications but does not include conceptual knowledge.
Semantics	The meanings that correspond to all lexical items and to all possible sentences.
Prosody	The vocal intonation that can modify the literal meaning of words and sentences.
Discourse	The linking of sentences such that they constitute a narrative.

A third set of structures, also located largely in the left hemisphere, mediates between the first two. It can take a concept and stimulate the production of word-forms, or it can receive words and cause the brain to evoke the corresponding concepts.

Such mediation structures have also been hypothesized from a purely psycholinguistic perspective. Willem J. M. Levelt of the Max Planck Institute for Psycholinguistics in Nijmegen has suggested that word-forms and sentences are generated from concepts by means of a component he calls "lemma," and Merrill F. Garret of the University of Arizona holds a similar view.

The concepts and words for colors serve as a particularly good example of this tripartite organization. Even those afflicted by congenital color blindness know that certain ranges of hue (chroma) band together and are different from other ranges, independent of their brightness and saturation. As Brent Berlin and Eleanor H. Rosch of the University of California at Berkeley have shown, these color concepts are fairly universal and develop whether or not a given culture actually has names to denote them. Naturally, the retina and the lateral geniculate nucleus perform the initial processing of color signals, but the primary visual cortex and at least two other cortical regions (known as V2 and V4) also participate in color processing; they fabricate what we know as the experience of color.

With our colleague Matthew Rizzo, we have found that damage to the occipital and subcalcarine portions of the left and right lingual gyri, the region of the brain believed to contain the V2 and V4 cortices, causes a condition called achromatopsia. Patients who previously had normal vision lose their perception of color. Furthermore, they lose the ability even to imagine colors. Achromatopsics usually see the world in shades of gray; when they conjure up a typically colored image in their minds, they see the shapes, movement, and texture but not the color. When they think about a field of grass, no green is available, nor will red or yellow be part of their otherwise normal evocation of blood or banana. No lesion elsewhere in the brain can cause a similar defect. In some sense, then, the concept of colors depends on this region.

Patients with lesions in the left posterior temporal and inferior parietal cortex do not lose access to their concepts, but they have a sweeping impairment of their ability to produce proper word morphology regardless of the category to which a word belongs. Even if they are properly experiencing a given color and attempting to retrieve the corresponding word-form, they produce phonemically distorted color names; they may say "buh" for "blue," for example.

Other patients, who sustain damage in the temporal segment of the left lingual gyrus, suffer from a peculiar defect called color anomia, which affects neither color concepts nor the utterance of color words. These patients continue to *experience* color normally: they can match different hues, correctly rank hues of different saturation, and easily put the correct colored paint chip next to objects in a black-and-white photograph. But their ability to put names to color is dismally impaired. Given the limited set of color names available to those of us who are not interior decorators, it is surprising to see patients use the word "blue" or "red" when shown green or yellow and yet be capable of neatly placing a green chip next to a picture of grass or a yellow chip next to a picture of a banana. The defect goes both ways: given a color name, the patient will point to the wrong color.

At the same time, however, all the wrong color names the patient uses are beautifully formed, phonologically speaking, and the patient has no other language impairment. The color-concept system is intact, and so is the word-form implementation system. The problem seems to reside with the neural system that mediates between the two.

The same three-part organization that explains how people manage to talk about color applies to other concepts as well. But how are such concepts physically represented in the brain? We believe there are no permanently held "pictorial" representations of objects or persons as was traditionally thought. Instead the brain holds, in effect, a record of the neural activity that takes place in the sensory and motor cortices during interaction with a given object. The records are patterns of synaptic connections that can re-create the separate sets of activity that define an object or event; each record can also stimulate related ones. For example, as a person picks up a coffee cup, her visual cortices will respond to the colors of the cup and of its contents as well as to its shape and position. The somatosensory cortices will register the shape the hand assumes as it holds the cup, the movement of the hand and the arm as they bring the cup to the lips, the warmth of the coffee, and the body change some people call pleasure when they drink the stuff. The brain does not merely represent aspects of external reality; it also records how the body explores the world and reacts to it.

The neural processes that describe the interaction between the individual and the object constitute a rapid sequence of microperceptions and microactions, almost simultaneous as far as consciousness is concerned. They occur in separate functional regions, and each region contains additional subdivisions: the visual aspect of perception, for example, is segregated within smaller systems specialized for color, shape, and movement.

Where can the records that bind together all these fragmented activities be held? We believe they are embodied in ensembles of neurons within the brain's many "convergence" regions. At these sites the axons of feedforward projecting neurons from one part of the brain converge and join with reciprocally diverging feedback projections from other regions. When reactivation within the convergence zones stimulates the feedback projections, many anatomically separate and widely distributed neuron ensembles fire simultaneously and reconstruct previous patterns of mental activity.

In addition to storing information about experiences with objects, the brain also categorizes the information so that related events and concepts—shapes, colors, trajectories in space and time, and pertinent body movements and reactions—can be reactivated together. Such categorizations are denoted by yet another record in another convergence zone. The essential properties of the entities and processes in any interaction are thus represented in an interwoven fashion. The collected knowledge that can be represented includes the fact that a coffee cup has dimensions and a boundary; that it is made of something and has parts; that if it is divided it no longer is a cup, unlike water, which retains its identity no matter how it is divided; that it moved along a particular path, starting at one point in space and ending at another; that arrival at its destination produced a specific outcome. These aspects of neural representation bear a strong resemblance to the primitives of conceptual structure proposed by Ray Jackendoff of Brandeis University and the cognitive semantic schemas hypothesized by George P. Lakoff of the University of California at Berkeley, both working from purely linguistic grounds.

Activity in such a network, then, can serve both understanding and expression. The activity in the network can reconstruct knowledge so that a person experiences it consciously, or it can activate a system that mediates between concept and language, causing appropriately correlated word-forms and syntactical structures to be generated. Because the brain categorizes perceptions and actions simultaneously along many different dimensions, symbolic representations such as metaphor can easily emerge from this architecture.

Damage to parts of the brain that participate in these neural patterns should produce cognitive defects that clearly delineate the categories according to which concepts are stored and retrieved (the damage that results in achromatopsia is but one example of many). Elizabeth K. Warrington of the National Hospital for Nervous Diseases in London has studied category-related recognition defects and found patients who lose cognizance of certain classes of object. Similarly, in collaboration with our colleague Daniel Tranel, we have shown that access to concepts in a number of domains depends on particular neural systems.

For example, one of our patients, known as Boswell, no longer retrieves concepts for any unique entity (a specific person, place, or event) with which he was previously familiar. He has also lost concepts for nonunique entities of particular classes. Many animals, for instance, are completely strange to him even though he retains the concept level that lets him know that they are living and animate. Faced with a picture of a raccoon he says, "It is an animal," but he has no idea of its size, habitat, or typical behavior.

Curiously, when it comes to other classes of nonunique entities, Boswell's cognition is apparently unimpaired. He can recognize and name objects, such as a wrench, that are manipulable and have a specific action attached to them. He can retrieve concepts for attributes of entities: he knows what it means for an object to be beautiful or ugly. He can grasp the idea of states or activities such as being in love, jumping, or swimming. And he can understand abstract relations among entities or events such as "above," "under," "into," "from," "before," "after," or "during." In brief, Boswell has an impairment of concepts for many entities, all of which are denoted by nouns (common and proper). He has no problem whatsoever with concepts for attributes, states, activities, and relations that are linguistically signified by adjectives, verbs, functors (prepositions, conjunctions, and other verbal connective tissue), and syntactic structures. Indeed, the syntax of his sentences is impeccable.

Lesions such as Boswell's, in the anterior and middle regions of both temporal lobes, impair the brain's conceptual system. Injuries to the left hemisphere in the vicinity of the sylvian fissure, in contrast, interfere with the proper formation of words and sentences. This brain system is the most thoroughly investigated of those involved in language. More than a century and a half ago Pierre Paul Broca and Carl Wernicke determined the rough location of these basic language centers and discovered the phenomenon known as cerebral dominance—in most humans language structures lie in the left hemisphere rather than the right. This disposition holds for roughly 99 percent of right-handed people and two thirds of left-handers. (The pace of research in this area has accelerated during the past two decades, thanks in large part to the influence of the late Norman Geschwind of Harvard Medical School and Harold Goodglass of the Boston Veterans Medical Center.)

Studies of aphasic patients (those who have lost part or all of their ability to speak) from different language backgrounds highlight the constancy of these structures. Indeed, Edward Klima of the University of California at San Diego and Ursula Bellugi of the Salk Institute for Biological Studies in San Diego have discovered that damage to the brain's word-formation systems is implicated in sign-language aphasia as well. Deaf individuals who suffer focal brain damage in the left hemisphere can lose the ability to sign or to understand sign language. Because the damage in question is not to the visual cortex, the ability to see signs is not in question, just the ability to interpret them.

In contrast, deaf people whose lesions lie in the right hemisphere, far from the regions responsible for word and sentence formation, may lose conscious awareness of objects on the left side of their visual field, or they may be unable to perceive correctly spatial relations among objects, but they do not lose the ability to sign or understand sign language. Thus, regardless of the sensory channel through which linguistic information passes, the left hemisphere is the base for linguistic implementation and mediation systems.

Investigators have mapped the neural systems most directly involved in word and sentence formations by studying the location of lesions in aphasic patients. In addition, George A. Ojemann of the University of Washington and Ronald P. Lesser and Barry Gordon of Johns Hopkins University have stimulated the exposed cerebral cortex of patients undergoing surgery for epilepsy and made direct electrophysiological recordings of the response.

Damage in the posterior perisylvian sector, for example, disrupts the assembly of phonemes into words and the selection of entire word-forms. Patients with such damage may fail to speak certain words, or they may form words improperly ("loliphant" for "elephant"). They may, in addition, substitute a pronoun or a word at a more general taxonomic level for a missing one ("people" for "woman") or use a word semantically related to the concept they intend to express ("headman" for "president"). Victoria A. Fromkin of the University of California at Los Angeles has elucidated many of the linguistic mechanisms underlying such errors.

Damage to this region, however, does not disrupt patients' speech rhythms or the rate at which they speak. The syntactic structure of sentences is undisturbed even when there are errors in the use of functor words such as pronouns and conjunctions.

Damage to this region also impairs processing of speech sounds, and so patients have difficulty understanding spoken words and sentences. Auditory comprehension fails not because, as has been traditionally believed, the posterior perisylvian sector is a center to store "meanings" of words but rather because the normal acoustic analyses of the word-forms the patient hears are aborted at an early stage.

The systems in this sector hold auditory and kinesthetic records of phonemes and the phoneme sequences that make up words. Reciprocal projections of neurons between the areas holding these records mean that activity in one can generate corresponding activity in the other.

These regions connect to the motor and premotor cortices, both directly and by means of a subcortical path that includes the left basal ganglia and nuclei in the forward portion of the left thalamus. This dual motor route is especially important: the actual production of speech sounds can take place under the control of either a cortical or a subcortical circuit, or both. The subcortical circuit corresponds to "habit learning," whereas the cortical route implies higher-level, more conscious control and "associative learning."

For instance, when a child learns the word-form "yellow," activations would pass through the word-formation and motor-control systems via both the cortical and subcortical routes, and activity in these areas would be correlated with the activity of the brain regions responsible for color concepts and mediation between concept and language. In time, we suspect, the concept-mediation system develops a direct route to the basal ganglia, and so the posterior perisylvian sector does not have to be strongly activated to produce the word "yellow." Subsequent learning of the word-form for yellow in another language would again require participation of the perisylvian region to establish auditory, kinesthetic, and motor correspondences of phonemes.

It is likely that both cortical "associative" and subcortical "habit" systems operate in parallel during language processing. One system or the other predominates depending on the history of language acquisition and the nature of the item. Steven Pinker of the Massachusetts Institute of Technology has suggested, for example, that most people acquire the past tense of irregular verbs (take, took, taken) by associative learning and that of regular verbs (those whose past tense ends in -ed) by habit learning.

The anterior perisylvian sector, on the front side of the rolandic fissure, appears to contain structures that are responsible for speech rhythms and grammar. The left basal ganglia are part and parcel of this sector, as they are of the posterior perisylvian one. The entire sector appears to be strongly associated with the cerebellum; both the basal ganglia and the cerebellum receive projections from a wide variety of sensory regions in the cortex and return projections to motor-related areas. The role of the cerebellum in language and cognition, however, remains to be elucidated.

Patients with damage in the anterior perisylvian sector speak in flat tones, with long pauses between words, and have defective grammar. They tend in particular to leave out conjunctions and pronouns, and grammatical order is often compromised. Nouns come easier to patients with these lesions than do verbs, suggesting that other regions are responsible for their production.

Patients with damage in this sector have difficulty understanding meaning that is conveyed by syntactic structures. Edgar B. Zurif of Brandeis University, Eleanor M. Saffran of Temple University, and Myrna F. Schwartz of Moss Rehabilitation Hospital in Philadelphia have shown that these patients do not always grasp reversible passive sentences such as "The boy was kissed by the girl," in which boy and girl are equally likely to be the recipient of the action. Nevertheless, they can still assign the correct meaning to a nonreversible passive sentence such as "The apple was eaten by the boy" or the active sentence "The boy kissed the girl."

The fact that damage to this sector impairs grammatical processing in both speech and understanding suggests that its neural systems supply the mechanics of component assembly at sentence level. The basal ganglia serve to assemble the components of complex motions into a smooth whole, and it seems reasonable that they might perform an analogous function in assembling word-forms into sentences. We also believe (based on experimental evidence of similar, although less extensive structures in monkeys) that these neural structures are closely interconnected with syntactic mediation units in the frontoparietal cortices of both hemispheres [see illustration on page 38]. The delineation of those units is a topic of future research.

Between the brain's concept-processing systems and those that generate words and sentences lie the mediation systems we propose. Evidence for this neural brokerage is beginning to emerge from the study of neurological patients. Mediation systems not only select the correct words to express a particular concept, but they also direct the generation of sentence structures that express relations among concepts.

When a person speaks, these systems govern those responsible for word formation and syntax; when a person understands speech the word-formation systems drive the mediation systems. Thus far we have begun to map the systems that mediate proper nouns and common nouns that denote entities of a particular class (for example, visually ambiguous, nonmanipulable entities such as most animals).

Consider the patients whom we will call A.N. and L.R., who had sustained damage to the anterior and midtemporal cortices. Both can retrieve concepts normally: when shown pictures of entities or substances from virtually any conceptual category—human faces, body parts, animals and botanical specimens, vehicles and buildings, tools and utensils—A.N. and L.R. know unequivocally what they are

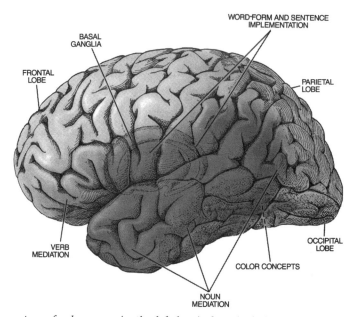

Brain systems for language in the left hemisphere include word and sentence-implementation structures and mediation structures for various lexical items and grammar. The collections of neural structures that represent the concepts themselves are distributed across both right and left hemispheres in many sensory and motor regions. (Carol Donner)

looking at. They can define an entity's functions, habitats, and value. If they are given sounds corresponding to those entities or substances (whenever a sound happens to be associated with them), A.N. and L.R. can recognize the items in question. They can perform this task even when they are blindfolded and asked to recognize an object placed in their hands.

But despite their obvious knowledge, they have difficulty in retrieving the names for many of the objects they know so well. Shown a picture of a raccoon, A.N. will say: "Oh! I know what it is—it is a nasty animal. It will come and rummage in your backyard and get into the garbage. The eyes and the rings in the tail give it away. I know it, but I cannot say the name." On the average they come up with less than half of the names they ought to retrieve. Their conceptual systems work well, but A.N. and L.R. cannot reliably access the word-forms that denote the objects they know.

The deficit in word-form retrieval depends on the conceptual category of the item that the patients are attempting to name. A.N. and L.R. make fewer errors for nouns that denote tools and utensils than for those naming animals, fruits, and vegetables. (This phenomenon has been reported in similar form by Warrington and her colleague Rosaleen A. McCarthy of the National Hospital for Nervous Diseases and by Alfonso Caramazza and his colleagues at Johns Hopkins University.) The patients' ability to find names, however, does not split neatly at the boundary of natural and man-made entities. A.N. and L.R. can produce the words for such natural stimuli as body parts perfectly, whereas they cannot do the same for musical instruments, which are as artificial and as manipulable as garden tools.

In brief, A.N. and L.R. have a problem with the retrieval of common nouns denoting certain entities regardless of their membership in particular conceptual categories. There are many reasons why some entities might be more or less vulnerable to lesions than others. Of necessity, the brain uses different neural systems to represent entities that differ in structure or behavior or entities that a person relates to in different ways.

A.N. and L.R. also have trouble with proper nouns. With few exceptions, they cannot name friends, relatives, celebrities, or places. Shown a picture of Marilyn Monroe, A.N. said, "Don't know her name but I know who she is; I saw her movies; she had an affair with the president; she committed suicide, or maybe somebody killed her; the police, maybe?" These patients do not have what is known as face agnosia or prosopagnosia—they can recognize a face without hesitation—but they simply cannot retrieve the word-form that goes with the person they recognize.

Curiously, these patients have no difficulty producing verbs. In experiments we conducted in collaboration with Tranel, these patients perform just as well as matched control subjects on tasks requiring them to generate a verb in response to more than 200 stimuli depicting diverse states and actions. They are also adept at the production of prepositions, conjunctions, and pronouns, and their sentences are well formed and grammatical. As they speak or write, they produce a narrative in which, instead of the missing noun, they will substitute words like "thing" or "stuff" or pronouns such as "it" or "she" or "they." But the verbs that animate the arguments of those sentences are properly selected and produced and properly marked with respect to tense and person. Their pronunciation and prosody (the intonation of the individual words and the entire sentence) are similarly unexceptionable.

The evidence that lexical mediation systems are confined to specific regions is convincing. Indeed, the neural structures that mediate between concepts and word-forms appear to be graded from back to front along the occipitotemporal axis of the brain. Mediation for many general concepts seems to occur at the rear, in the more posterior left temporal regions; mediation for the most specific concepts takes place at the front, near the left temporal pole. We have now seen many patients who have lost their proper nouns but retain all or most of their common nouns. Their lesions are restricted to the left temporal pole and medial temporal surface of the brain, sparing the lateral and inferior temporal lobes. The last two, in contrast, are always damaged in the patients with common noun retrieval defects.

Patients such as A.N. and L.R., whose damage extends to the interior and midtemporal cortices, miss many common nouns but still name colors quickly and correctly. These correlations between lesions and linguistic defects indicate that the temporal segment of the left lingual gyrus supports mediation between color concepts and color names, whereas mediation between concepts for unique persons and their corresponding names requires neural structures at the opposite end of the network, in the left anterior temporal lobe. Finally, one of our more recent patients, G.J., has extensive damage that encompasses all of these left occipitotemporal regions from front to back. He has lost access to a sweeping universe of noun word-forms and is equally unable to name colors or unique persons. And yet his concepts are preserved. The results in these patients support Ojemann's finding of impaired language processing after electrical stimulation of cortices outside the classic language areas.

It appears that we have begun to understand fairly well where nouns are mediated, but where are the verbs? Clearly, if patients such as A.N. and L.R. can retrieve verbs and functor words normally, the regions required for those parts of speech cannot be in the left temporal region. Preliminary evidence points to frontal and parietal sites. Aphasia studies performed by our group and by Caramazza and Gabriele Miceli of Catholic University of the Sacred Heart, Milan, and Rita Berndt of the University of Maryland show that patients with left frontal damage have far more trouble with verb retrieval than with noun retrieval.

Additional indirect evidence comes from positron emission tomography (PET) studies conducted by Steven E. Petersen, Michael I. Posner, and Marcus E. Raichle of Washington University. They asked research subjects to generate a verb corresponding to the picture of an object—for example, a picture of an apple might generate "eat." These subjects activated a region of the lateral and inferior dorsal frontal cortex that corresponds roughly to the areas delineated in our studies. Damage to these regions not only compromises access to verbs and functors, it also disturbs the grammatical structure of the sentences that patients produce.

Although this phenomenon may seem surprising at first, verbs and functors constitute the core of syntactic structure, and so it makes sense that the mediation systems of syntax would overlap with them. Further investigations, either of aphasic patients or of normal subjects, whose brain activity can be mapped by PET scans, may clarify the precise arrangement of these systems and yield maps like those that we have produced to show the differing locations of common and proper nouns.

During the past two decades, progress in understanding the brain structures responsible for language has accelerated significantly. Tools such as magnetic resonance imaging have made it possible to locate brain lesions accurately in patients suffering from aphasia and to correlate specific language deficits with damage to particular regions of the brain. And PET scans offer the opportunity to study the brain activity of normal subjects engaged in linguistic tasks.

Considering the profound complexity of language phenomena, some may wonder whether the neural machinery that allows it all to happen will ever be understood. Many questions remain to be answered about how the brain stores concepts. Mediation systems for parts of speech other than nouns, verbs and functors, have been only partially explored. Even the structures that form words and sentences, which have been under study since the middle of the 19th century, are only sketchily understood.

Nevertheless, given the recent strides that have been made, we believe these structures will eventually be mapped and understood. The question is not if but when.

—September 1992

Visualizing the Mind

*Strategies of cognitive science and techniques
of modern brain imaging open a window
to the neural systems responsible for thought*

Marcus E. Raichle

What causes the pity we might feel for the melancholy Dane in *Hamlet* or the chill during a perusal of "The Raven"? Our brains have absorbed from our senses a printed sequence of letters and then converted them into vivid mental experiences and potent emotions. The "black box" description of the brain, however, fails to pinpoint the specific neural processes responsible for such mental actions. While philosophers have for centuries pondered this relation between mind and brain, investigators have only recently been able to explore the connection analytically—to peer inside the black box. The ability stems from developments in imaging technology that the past few years have seen, most notably positron emission tomography and magnetic resonance imaging. Coupled with powerful computers, these techniques can now capture, in real time, images of the physiology associated with thought processes. They show how specific regions of the brain "light up" when activities such as reading are performed and how neurons and their elaborate cast of supporting cells organize and coordinate their tasks. The mapping of thought can also act as a tool for neurosurgery and elucidate the neural differences of people crippled by devastating mental illnesses, including depression and schizophrenia.

I hasten to point out that the underlying assumptions of current brain mapping are distinct from those held by early phrenologists. They posited that single areas of the brain, often identified by bumps on the skull, uniquely represented specific thought processes and emotions. In contrast, modern thinking posits that networks of neurons residing in strictly localized areas perform thought processes. So just as specific members of a large orchestra perform together in a precise fashion to produce a symphony, a group of localized brain areas performing elementary operations work together to exhibit an observable human behavior. The foundation for such analyses is that complex behaviors can be broken down into a set of constituent mental operations. In order to read, for example, one must recognize that a string of letters is a word; then recognize the meaning of words, phrases, or sentences; and finally create mental images.

The challenge, of course, is to determine those parts of the brain that are active and those that are dormant during the performance of tasks. In the past, cognitive neuroscientists have relied on studies of laboratory animals and patients with localized brain injuries to gain insight into the brain's functions. Imaging techniques, however, permit us to visualize safely the anatomy and the function of the normal human brain.

The modern era of medical imaging began in the early 1970s, when the world was introduced to a remarkable technique called x-ray computed tomography, now known as x-ray CT, or just CT. South African physicist Allan M. Cormack and British engineer Sir Godfrey Hounsfield independently developed its principles. Hounsfield constructed the first CT instrument in England. Both investigators received the Nobel Prize in 1979 for their contributions.

Computed tomography takes advantage of the fact that different tissues absorb varying amounts of x-ray energy. The denser the tissue, the more it absorbs. A highly focused beam of x-rays traversing through the body will exit at a reduced level depending on the tissues and organs through which it passed. A beam of x-rays passed through the body at many different angles through a plane collects sufficient information to reconstruct a picture of the body section. Crucial in the development of x-ray CT was the emergence of clever computing and mathematical techniques to process the vast amount of information necessary to create images themselves. Without the availability of sophisticated computers, the task would have been impossible to accomplish.

X-ray CT had two consequences. First, it changed forever the practice of medicine because it was much superior to standard x-rays. For the first time, investigators could safely and effectively view living human tissue such as the brain with no discomfort to the patient. Standard x-rays revealed only bone and some surrounding soft tissue. Second, it immediately stimulated scientists and engineers to consider alternative ways of creating images of the body's interior using similar mathematical and computer strategies for image reconstruction.

One of the first such groups to be intrigued by the possibilities opened by computed tomography consisted of experts in tissue autoradiography, a method used for many years in animal studies to investigate organ metabolism and blood flow. In tissue autoradiography, a radioactively labeled compound is injected into a vein. After the compound has accumulated in the organ (such as the brain) under interest, the animal is sacrificed and the organ removed for study. The organ is carefully sectioned, and the individual slices are laid on a piece of film sensitive to radioactivity. Much as the film in a camera records a scene as you originally viewed it, this x-ray film records the distribution of radioactively labeled compound in each slice of tissue.

Once the x-ray film is developed, scientists have a picture of the distribution of radioactivity within the organ and hence can deduce the organ's specific functions. The type of information is determined by the radioactive compound injected. A radioactively labeled form of glucose, for example, measures brain metabolism because glucose is the primary source of energy for neurons. Louis Sokoloff of the National Institute of Mental Health introduced this now widely used autoradiographic method in 1977.

Investigators adept with tissue autoradiography became fascinated when CT was introduced. They suddenly realized that if they could reconstruct the anatomy of an organ by passing an x-ray beam through it, they could also reconstruct the distribution of a previously administered radioisotope. One had simply to measure the emission of radioactivity from the body section. With this realization was born the idea of autoradiography of living human subjects.

A crucial element in the evolution of human autoradiography was the choice of radioisotope. Workers in the field selected a class of radioisotopes that emit positrons, which resemble electrons except that they carry a positive charge. A positron would almost immediately combine with a nearby electron. They would annihilate each other, emitting two gamma rays in the process. Because each gamma ray travels in nearly opposite directions, devices around the sample would detect the gamma rays and locate their origin. The crucial role of positrons in human autoradiography gave rise to the name positron emission tomography, or PET.

Throughout the late 1970s and early 1980s, researchers rapidly developed PET to measure various activities in the brain, such as glucose metabolism, oxygen consumption, blood flow, and interactions with drugs. Of these variables, blood flow has proved the most reliable indicator of moment-to-moment brain function.

The idea that local blood flow is intimately related to brain function is a surprisingly old one. English physiologists Charles S. Roy and Charles S. Sherrington formally presented the idea in a publication in 1890. They suggested that an "automatic mechanism" regulated the blood supply to the brain. The amount of blood depended on local variations in activity. Although subsequent experiments have amply confirmed the existence of such an automatic mechanism, no one as yet is entirely certain about its exact nature. It obviously remains a challenging area for research.

PET measures blood flow in the normal human brain by adapting an autoradiographic technique for laboratory animals developed in the late 1940s by Seymour S. Kety of the National Institute of Mental Health and his colleagues. PET relies on radioactively labeled water—specifically, hydrogen combined with oxygen 15, a radioactive isotope of oxygen. The labeled water emits copious numbers of positrons as it decays (hydrogen isotopes cannot be used, because none emit positrons). The labeled water is administered into a vein in the arm. In just over a minute the radioactive water accumulates in the brain, forming an image of blood flow.

The radioactivity of the water produces no deleterious effects. Oxygen 15 has a half-life of only two minutes; an entire sample decays almost completely in about 10 minutes (five half-lives) into a nonradioactive form. The rapid decay substantially reduces the exposure of subjects to the potentially harmful effects of radiation. Moreover, only low doses of the radioactive label are necessary.

The fast decay and small amounts permit many measurements of blood flow to be made in a single experiment. In this way, PET can take multiple pictures of the brain at work. Each picture serves as a snapshot capturing the momentary activity within the brain. Typical PET systems can locate changes in activity with an accuracy of a few millimeters.

A distinct strategy for the functional mapping of neuronal activity by PET has emerged during the past 10 years. This approach extends an idea first introduced to psychology in 1868 by Dutch physiologist Franciscus C. Donders. Donders proposed a general method to measure thought processes based on a simple logic. He subtracted the time needed to respond to a light (with, say, the press of a key) from the time needed to respond to a particular color of light. He found that discriminating color required about 50 milliseconds. In this way, Donders isolated and measured a mental process for the first time.

The current PET strategy is designed to accomplish a similar subtraction but in terms of the brain areas implementing the mental process. In particular, images of blood flow taken before a task is begun are compared with those obtained when the brain is engaged in that task. Investigators refer to these two periods as the control state and the task state. Workers carefully choose each state so as to isolate as best as possible a limited number of mental operations. Subtracting blood-flow measurements made in the control state from each task state indicates those parts of the brain active during a particular task.

To achieve reliable data, workers take the average of responses across many individual subjects or of many experimental trials in the same person. Averaging enables researchers to detect changes in blood flow associated with mental activity that would otherwise be easily confused with spurious shifts resulting from noise.

One of the first assignments in which PET blood-flow mapping has proved useful is in the study of language. The manner in which language skills are acquired and organized in the human brain has been the subject of intense investigation for more than a century. Work began in earnest in 1861, when French physician Pierre Paul Broca described a patient whose damaged left frontal lobe destroyed the ability to speak. (To this day, patients who have frontal lobe damage and have trouble speaking are often referred to as having Broca's aphasia.) Broca's studies of language localization were complemented by Carl Wernicke, a German neurologist. In 1874 Wernicke told of people who had difficulty comprehending language. They harbored damage to the left temporal lobe, a region now usually referred to as Wernicke's area. From these beginnings has emerged a concept of language organization in the human brain: information flows from visual and auditory reception to areas in the left temporal lobe for comprehension and then on to frontal areas for speech production.

All this information was gleaned from brain-damaged patients. Can investigators derive insight about language organization from a healthy brain? In 1988 my colleagues Steven E. Petersen, Michael I. Posner, Peter T. Fox, and Mark A. Mintun and I at the Washington University Medical Center began a series of studies to answer just this question. The initial study was based on a PET analysis of a seemingly simple job: speaking an appropriate verb when presented with a common English noun. For example, a subject might see or hear the word "hammer," to which an appropriate response might be "hit."

We chose this assignment because it could be broken down into many components. Each component could separately be analyzed through a careful selection of tasks. The most readily apparent elements include visual and auditory word perception, the organization and execution for word output (speech), and the processes by which the brain retrieves the meanings of words. (Of course, each of these operations can be divided further into several additional subcomponents.)

To identify the areas of the brain used in a particular operation we composed four levels of information processing. Such a hierarchy has become standard among laboratories doing this type of research [see color plate 2]. In the first level, subjects were asked to fix their gaze on a pair of small crosshairs—the arrangement looks like a small plus sign—in the middle of a television monitor. At the same time a PET scan measured blood flow in the brain, providing a snapshot of mental activity.

In the second level, subjects continued to maintain their gaze on the crosshairs as blood flow was measured, but during this scan they were exposed to common English nouns. The nouns either appeared below the crosshairs on the television monitor or were spoken through earphones (separate scans were performed for visual and auditory presentations). In the third level, subjects were asked to recite the word they viewed or heard. Finally, in the fourth level, the subjects said out loud a verb appropriate for the noun.

Subtracting the first level from the second isolated those brain areas concerned with visual and auditory word perception. Deducting the second level from the third pinpointed those parts of the brain concerned with speech production. Subtracting level three from level four located those regions concerned with selecting the appropriate verb to a presented noun.

The final subtraction (speaking nouns minus generating verbs) was of particular interest, because it provided a portrait of pure mental activity (perception and speech—or input and output—having been subtracted away). This image permitted us to view what occurs in our brains as we interpret the meaning of words and, in turn, express meaning through their use. It renders visible conscious function because much of our thinking is carried out by concepts and ideas represented by words.

The results of this study clearly demonstrate how brain imaging can relate mental operations of a behavioral task to specific networks of brain areas orchestrated to perform each operation. As anticipated by cognitive scientists and neuroscientists, the apparently simple task of generating a verb for a presented noun is not accomplished by a single part of the brain but rather by many areas organized into networks. Perception of visually presented words occurs in a network of areas in the back of the brain, where many components of the brain's visual system reside. Perception of aurally presented words occurs in an entirely separate network of areas—in our temporal lobes.

Speech production (that is, simply repeating out loud the presented nouns) predictably involves motor areas of the brain. Regions thought to be Broca's and Wernicke's areas do not appear to be engaged routinely in this type of speech production, an activity that would be viewed by many as quite automatic for most fluent speakers in their native language. This finding suggests what we might have suspected: we occasionally speak without consciously thinking about the consequences.

Regions of the left frontal and temporal lobes (those corresponding in general to the respective locations of Broca's and Wernicke's areas) only become active when two tasks are added: consciously assessing word meaning and choosing an appropriate response. Moreover, two other areas come into play under these circumstances, forming a network of four brain regions. Interestingly, two areas used in the routine repetition of words were turned off. This shutdown suggests that the demands of generating a verb into a presented noun does not simply build on the task of just saying the noun. Rather the act of speaking a verb to a presented noun differs from speaking the noun, as far as the brain is concerned.

This finding caused us to pause and consider what would happen if we allowed subjects a few minutes of practice on their task of generating verbs. Although subjects initially discover that forming verbs rapidly is difficult (nouns are presented every 1.5 seconds), they become relaxed and proficient after 15 minutes of practice. An examination of the brain after training reveals that practice completely changes the neural circuits recruited [*see color plate 2*]. The circuits responsible for noun repetition now generate the verbs. Thus, practice not only makes perfect (something we have always known) but also changes the way our brain organizes itself (something we may not have fully appreciated).

As cognitive neuroscientists demonstrated the utility of PET technology, a newer method swiftly emerged that could compete with PET's abilities. Magnetic resonance imaging, or MRI, has now become a fairly common tool for diagnosing tissue damage. Recent developments have vastly increased the speed with which MRI can form images, thus making it suitable for research in cognitive neuroscience.

MRI derives from a potent laboratory technique known as nuclear magnetic resonance (NMR), which was designed to explore detailed chemical features of molecules. It garnered a Nobel Prize for its developers, Felix Bloch of Stanford University and Edward M. Purcell of Harvard University, in 1952. The method depends on the fact that many atoms behave as little compass needles in the presence of a magnetic field. By skillfully manipulating the magnetic field, scientists can align the atoms. Applying radio-wave pulses to the sample under these conditions perturbs the atoms in a precise manner. As a result, they emit detectable radio signals unique to the number and state of the particular atoms in the sample. Careful adjustments to the magnetic field and the radio-wave pulses yield particular information about the sample under study.

NMR moved from the laboratory to the clinic when Paul C. Lauterbur of the University of Illinois found that NMR can form images by detecting protons. Protons are useful because they are abundant in the human body and, by acting as little compass needles, respond sensitively to magnetic fields. Their application resulted in excellent images of the anatomy of organs that far surpassed in detail those produced by x-ray CT. Because the term "nuclear" made the procedure sound dangerous, NMR soon became known as magnetic resonance imaging.

The current excitement over MRI for brain imaging stems from the technique's ability to detect a signal inaccessible to PET scans. Specifically, it can detect an increase in oxygen that occurs in an area of heightened neuronal activity. The basis for this capacity comes from the way neurons make use of oxygen. PET scans had revealed that functionally induced increases in blood flow accompanied alterations in the amount of glucose the brain consumed but not in the amount of oxygen it used. In effect, the normal human brain during spurts of neuronal activity resorts to anaerobic metabolism. Few had suspected that the brain might rely on tactics similar to those used by sprinters' muscles. In fact, this form of metabolism occurs despite the presence of abundant oxygen in the normal brain. Why the brain acts this way is a mystery worthy of intense scientific scrutiny.

Additional blood to the brain without a concomitant increase in oxygen consumption leads to a heightened concentration of oxygen in the small veins draining the active neural centers. The reason is that supply has increased, but the demand has not. Therefore, the extra oxygen delivered to the active part of brain simply returns to the general circulation by way of the draining veins.

Why does oxygen play a crucial role in MRI studies of the brain? The answer lies in a discovery made by Nobel laureate Linus C. Pauling in 1935. He found that the amount of oxygen carried by hemoglobin (the molecule that transports oxygen and gives blood its red color) affects the magnetic properties of the hemoglobin. In 1990 Seiji Ogawa and his colleagues at AT&T Bell Laboratories demonstrated that MRI could detect these small magnetic fluctuations. Several research groups immediately realized the importance of this observation. By the middle of 1991 investigators showed that MRI can detect the functionally induced changes in blood oxygenation in the human brain. The ability of MRI machines to detect functionally induced changes in blood oxygenation leads many to refer to the technique as functional MRI, or fMRI.

Functional MRI has several advantages over x-ray CT and other imaging techniques. First, the signal comes directly from functionally induced changes in the brain tissue (that is, the change in venous oxygen concentration). Nothing, radioactive or otherwise, needs to be injected to obtain a signal. Second, MRI provides both anatomical and functional information in each subject, hence permitting an accurate structural identification of the active regions. Third, the spatial resolution is quite good, distinguishing parts as small as one to two millimeters (better than PET's resolution). Fourth, when properly equipped (that is, given so-called echoplanar capability), MRI can monitor the rate of change in the blood-flow-induced oxygen signal in real time.

Finally, MRI has little, if any, known biological risk. Some workers have raised concerns about the intensity of the magnetic field to which the tissues are exposed. So far most studies have found the effects to be benign. The largest drawback is the

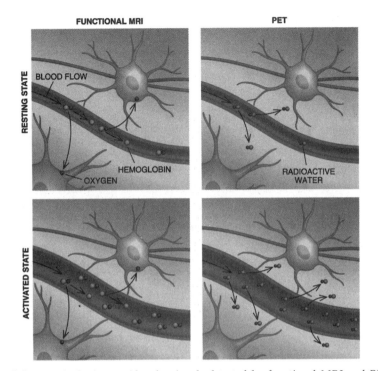

Blood flow to the brain provides the signals detected by functional MRI and PET. When resting neurons (top) become active (bottom), blood flow to them increases. MRI (left) detects changes in oxygen levels, which rise in the nearby blood vessels because active neurons consume no more oxygen than when they are at rest. PET (right) relies on the increased delivery of injected radioactive water, which diffuses out of the vessels to reach all parts of the brain. (Guilbert Gates)

claustrophobia some subjects may suffer. In most instrument designs the entire body must be inserted into a relatively narrow tube.

Several intriguing results with functional MRI were reported this past year. Robert G. Shulman and his colleagues at Yale University have confirmed PET findings about language organization in the brain. Using conventional, hospital-based MRI, Walter Schneider and Jonathan D. Cohen and their colleagues at the University of Pittsburgh have corroborated work in monkeys that indicated the primate visual cortex is organized into topographic maps that reflect the spatial organization of the world as we see it. Other groups are actively trying to visualize other forms of mental activity, such as the way we create mental images and memories.

The ability of MRI systems to monitor the oxygen signal in real time has suggested to some the possibility of measuring the time it takes for different brain areas to exchange information. Conceptually, one might think of individuals in the midst of a conference call. The temporal information sought would be equivalent to knowing who was speaking when and, possibly, who was in charge. Such information would be critical in understanding how specific brain areas coordinate as a network to produce behavior.

The stumbling block, however, is the speed of neuronal activity compared with the rate of change of oxygenation levels. Signals from one part of the brain can travel to another in 0.01 second or less. Unfortunately, changes in blood flow and blood oxygenation are much slower, occurring hundreds of milliseconds to several seconds later. MRI would not be able to keep up with the "conversations" between brain areas. The only methods that respond quickly enough are electrical recording techniques. Such approaches include electroencephalography (EEG), which detects brain electrical activity from the scalp, and magnetoencephalography (MEG), which measures the magnetic fields generated by electrical activity within the brain.

Why don't researchers just use EEG or MEG for the whole job of mapping brain function? The limitations are spatial resolution and sensitivity. Even though great strides in resolution have been made, especially with MEG, accurate localization of the source of brain activity remains difficult with electrical recording devices. Furthermore, the resolution becomes poorer the deeper into the brain we attempt to image.

Neither MRI nor PET suffers from this difficulty. They both can sample all parts of the brain with equal spatial resolution and sensitivity. As a result, a collaboration seems to be in the making between PET and MRI and electrical recording. PET and MRI, working in a combination yet to be determined, can define the anatomy of the circuits underlying a behavior of interest; electrical recording techniques can reveal the course of temporal events in these spatially defined circuits.

Regardless of the particular mix of technologies that will ultimately be used to image human brain function, the field demands extraordinary resources. Expensive equipment dominates this work. MRI, PET, and MEG equipment costs from $2 million to $4 million and is expensive to maintain. Furthermore, success requires close collaboration within multidisciplinary teams of scientists and engineers working daily with these tools. Institutions fortunate enough to have the necessary technical and human resources need to make them available to scientists at institutions less fortunate. Although some radiology departments have such equipment, the devices are usually committed mostly for patient care.

In addition to the images of brain activity, the experiments provide a vast amount of information. Such an accumulation not only yields answers to the questions posed at the time of the experiment but also provides invaluable information for future research, as those of us in the field have repeatedly discovered to our amazement and delight. Recent efforts to create neuroscience databases could organize and quickly disseminate such a repository of information.

Wise use of these powerful new tools and the data they produce can aid our understanding and care of people who have problems ranging from developmental learning disorders to language disabilities rising from, say, stroke. Researchers have begun to use functional brain imaging to learn about the mood disturbances that afflict patients with such mental illnesses as depression. The technology could guide neurosurgeons in the excision of brain tumors, enabling them to judge how the removal of tissue will hamper the patient. Centers across the world are investigating such other mental activities as attention, memory, perception, motor control, and emotion. Clearly, we are headed toward a much richer grasp of the relation between the human mind and the brain.

—April 1994

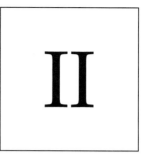

Reasoning
and Intelligence

THE GENERAL
INTELLIGENCE FACTOR

*Despite some popular assertions, a single factor
for intelligence, called g, can be measured
with IQ tests and does predict success in life*

Linda S. Gottfredson

No subject in psychology has provoked more intense public controversy than the study of human intelligence. From its beginning, research on how and why people differ in overall mental ability has fallen prey to political and social agendas that obscure or distort even the most well-established scientific findings. Journalists, too, often present a view of intelligence research that is exactly the opposite of what most intelligence experts believe. For these and other reasons, public understanding of intelligence falls far short of public concern about it. The IQ experts discussing their work in the public arena can feel as though they have fallen down the rabbit hole into Alice's Wonderland.

The debate over intelligence and intelligence testing focuses on the question of whether it is useful or meaningful to evaluate people according to a single major dimension of cognitive competence. Is there indeed a general mental ability we commonly call "intelligence," and is it important in the practical affairs of life? The answer, based on decades of intelligence research, is an unequivocal yes. No matter their form or content, tests of mental skills invariably point to the existence of a global factor that permeates all aspects of cognition. And this factor seems to have considerable influence on a person's practical quality of life. Intelligence as measured by IQ tests is the single most effective predictor known of individual performance at school and on the job. It also predicts many other aspects of well-being, including a person's chances of divorcing, dropping out of high school, being unemployed, or having illegitimate children.

By now the vast majority of intelligence researchers take these findings for granted. Yet in the press and in public debate, the facts are typically dismissed, downplayed, or ignored. This misrepresentation reflects a clash between a deeply felt ideal and a stubborn reality. The ideal, implicit in many popular critiques of intelligence research, is that all people are born equally able and that social inequality results only from the exercise of unjust privilege. The reality is that Mother Nature is no egalitarian. People are in fact unequal in intellectual potential—and they are born that way, just as they are born with different potentials for height, physical attractiveness, artistic flair, athletic prowess, and other traits. Although subsequent experience shapes this potential, no amount of social engineering can make individuals with widely divergent mental aptitudes into intellectual equals.

Of course, there are many kinds of talent, many kinds of mental ability, and many other aspects of personality and character that influence a person's chances of happiness and success. The functional importance of general mental ability in everyday life, however, means that without onerous restrictions on individual liberty, differences in mental competence are likely to result in social inequality. This gulf between equal opportunity and equal outcomes is perhaps what pains Americans most about the subject of intelligence. The public intuitively knows what is at stake: when asked to rank personal qualities in order of desirability, people put intelligence second only to good health. But with a more realistic approach to the intellectual differences between people, society could better accommodate these differences and minimize the inequalities they create.

■ EXTRACTING *g*

Early in the century-old study of intelligence, researchers discovered that all tests of mental ability ranked individuals in about the same way. Although mental tests are often designed to measure specific domains of cognition—verbal fluency, say, or mathematical skill, spatial visualization, or memory—people who do well on one kind of test tend to do well on the others, and people who do poorly generally do so across the board. This overlap, or intercorrelation, suggests that all such tests measure some global element of intellectual ability as well as specific cognitive skills. In recent decades, psychologists have devoted much effort to isolating that general factor, which is abbreviated g, from the other aspects of cognitive ability gauged in mental tests.

The statistical extraction of g is performed by a technique called factor analysis. Introduced at the turn of the century by British psychologist Charles Spearman, factor analysis determines the minimum number of underlying dimensions necessary to explain a pattern of correlations among measurements. A general factor suffusing all tests is not, as is sometimes argued, a necessary outcome of factor analysis. No general factor has been found in the analysis of personality tests, for example; instead the method usually yields at least five dimensions (neuroticism, extraversion, conscientiousness, agreeableness, and openness to ideas), each relating to different subsets of tests. But, as Spearman observed, a general factor does emerge from analysis of mental ability tests, and leading psychologists, such as Arthur R. Jensen of the University of California at Berkeley and John B. Carroll of the University of North Carolina at Chapel Hill, have confirmed his findings in the decades since. Partly because of this research, most intelligence experts now use g as the working definition of intelligence.

The general factor explains most differences among individuals in performance on diverse mental tests. This is true regardless of what specific ability a test is meant to assess, regardless of the test's manifest content (whether words, numbers, or figures), and regardless of the way the test is administered (in written or oral form, to an individual or to a group). Tests of specific mental abilities do measure those abilities, but they all reflect g to varying degrees as well. Hence, the g factor can be extracted from scores on any diverse battery of tests.

Conversely, because every mental test is "contaminated" by the effects of specific mental skills, no single test measures only g. Even the scores from IQ tests—which usually combine about a dozen subtests of specific cognitive skills—contain some "impurities" that reflect those narrower skills. For most purposes, these impurities make no practical difference, and g and IQ can be used interchangeably. But if they need to, intelligence researchers can statistically separate the g component of IQ. The ability to isolate g has revolutionized research on general intelligence, because it has allowed investigators to show that the predictive value of mental tests derives almost entirely from this global factor rather than from the more specific aptitudes measured by intelligence tests.

In addition to quantifying individual differences, tests of mental abilities have also offered insight into the meaning of intelligence in everyday life. Some tests and test items are known to correlate better with g than others do. In these items the "active

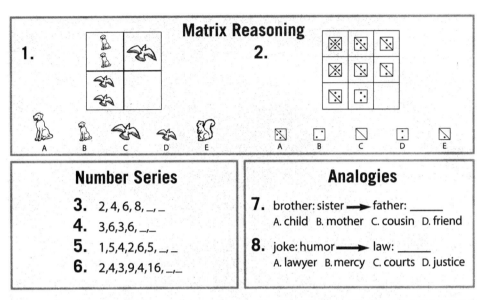

Answers: 1. A; 2. D; 3. 10, 12; 4. 3, 6; 5. 3, 7; 6. 5, 25; 7. B; 8. D

Sample IQ items resembling those on current tests require the test taker to fill in the empty spaces based on the pattern in the images, numbers, or words. Because they can vary in complexity, such tasks are useful in assessing g level. (Linda S. Gottfredson)

ingredient" that demands the exercise of g seems to be complexity. More complex tasks require more mental manipulation, and this manipulation of information—discerning similarities and inconsistencies, drawing inferences, grasping new concepts and so on—constitutes intelligence in action. Indeed, intelligence can best be described as the ability to deal with cognitive complexity.

This description coincides well with lay perceptions of intelligence. The g factor is especially important in just the kind of behaviors that people usually associate with "smarts": reasoning, problem solving, abstract thinking, quick learning. And whereas g itself describes mental aptitude rather than accumulated knowledge, a person's store of knowledge tends to correspond with his or her g level, probably because that accumulation represents a previous adeptness in learning and in understanding new information. The g factor is also the one attribute that best distinguishes among persons considered gifted, average, or retarded.

Several decades of factor-analytic research on mental tests have confirmed a hierarchical model of mental abilities. The evidence, summarized most effectively in Carroll's 1993 book, *Human Cognitive Abilities,* puts g at the apex in this model, with more specific aptitudes arrayed at successively lower levels: the so-called group factors, such as verbal ability, mathematical reasoning, spatial visualization, and memory, are just below *g,* and below these are skills that are more dependent on knowledge or experience, such as the principles and practices of a particular job or profession.

Some researchers use the term "multiple intelligences" to label these sets of narrow capabilities and achievements. Psychologist Howard Gardner of Harvard University, for example, has postulated that eight relatively autonomous "intelligences" are exhibited in different domains of achievement. He does not dispute the existence of *g* but treats it as a specific factor relevant chiefly to academic achievement and to situations that resemble those of school. Gardner does not believe that tests can fruitfully measure his proposed intelligences; without tests, no one can at present determine whether the intelligences are indeed independent of *g* (or each other). Furthermore, it is not clear to what extent Gardner's intelligences tap personality traits or motor skills rather than mental aptitudes.

Other forms of intelligence have been proposed; among them, emotional intelligence and practical intelligence are perhaps the best known. They are probably amalgams either of intellect and personality or of intellect and informal experience in specific job or life settings, respectively. Practical intelligence like "street smarts," for example, seems to consist of the localized knowledge and know-how developed with untutored experience in particular everyday settings and activities—the so-called school of hard knocks. In contrast, general intelligence is not a form of achievement, whether local or renowned. Instead the *g* factor regulates the rate of learning: it greatly affects the rate of return in knowledge to instruction and experience but cannot substitute for either.

■ THE BIOLOGY OF *g*

Some critics of intelligence research maintain that the notion of general intelligence is illusory; that no such global mental capacity exists and that apparent "intelligence" is really just a by-product of one's opportunities to learn skills and information valued in a particular cultural context. True, the concept of intelligence and the way in which individuals are ranked according to this criterion could be social artifacts. But the fact that *g* is not specific to any particular domain of knowledge or mental skill suggests that *g* is independent of cultural content, including beliefs about what intelligence is. And tests of different social groups reveal the same continuum of general intelligence. This observation suggests either that cultures do not construct *g* or that they construct the same *g.* Both conclusions undercut the social artifact theory of intelligence.

Moreover, research on the physiology and genetics of *g* has uncovered biological correlates of this psychological phenomenon. In the past decade, studies by teams of researchers in North America and Europe have linked several attributes of the brain to general intelligence. After taking into account gender and physical stature, brain size as determined by magnetic resonance imaging is moderately correlated with IQ (about 0.4 on a scale of 0 to 1). So is the speed of nerve conduction. The brains of bright people also use less energy during problem solving than do those of their less able peers. And various qualities of brain waves correlate strongly (about 0.5 to 0.7) with IQ: the brain waves of individuals with higher IQs, for example, respond more promptly and consistently to simple sensory stimuli such as audible clicks. These observations have led some investigators to posit that differences in g result from differences in the speed and efficiency of neural processing. If this theory is true, environmental conditions could influence g by modifying brain physiology in some manner.

Studies of so-called elementary cognitive tasks (ECTs), conducted by Jensen and others, are bridging the gap between the psychological and the physiological aspects of g. These mental tasks have no obvious intellectual content and are so simple that adults and most children can do them accurately in less than a second. In the most basic reaction-time testers, for example, the subject must react when a light goes on by lifting her index finger off a home button and immediately depressing a response button. Two measurements are taken: the number of milliseconds between the illumination of the light and the subject's release of the home button, which is called decision time, and the number of milliseconds between the subject's release of the home button and pressing of the response button, which is called movement time.

In this task, movement time seems independent of intelligence, but the decision times of higher-IQ subjects are slightly faster than those of people with lower IQs. As the tasks are made more complex, correlations between average decision times and IQ increase. These results further support the notion that intelligence equips individuals to deal with complexity and that its influence is greater in complex tasks than in simple ones.

The ECT-IQ correlations are comparable for all IQ levels, ages, genders, and racial-ethnic groups tested. Moreover, studies by Philip A. Vernon of the University of Western Ontario and others have shown that the ECT-IQ overlap results almost entirely from the common g factor in both measures. Reaction times do not reflect differences in motivation or strategy or the tendency of some individuals to rush through tests and daily tasks—that penchant is a personality trait. They actually seem to measure the speed with which the brain apprehends, integrates, and evaluates information. Research on ECTs and brain physiology has not yet identified the biological determinants of this processing speed. These studies do suggest, however, that g is as reliable and global a phenomenon at the neural level as it is at the level of the complex information processing required by IQ tests and everyday life.

The existence of biological correlates of intelligence does not necessarily mean that intelligence is dictated by genes. Decades of genetics research have shown, however, that people are born with different hereditary potentials for intelligence and that these genetic endowments are responsible for much of the variation in mental ability among individuals. An international team of scientists headed by Robert Plomin of the Institute of Psychiatry in London announced the discovery of the first gene linked to intelligence. Of course, genes have their effects only in interaction with environments, partly by enhancing an individual's exposure or sensitivity to formative experiences. Differences in general intelligence, whether measured as IQ or, more accurately, as g are both genetic and environmental in origin—just as are all other psychological traits and attitudes studied so far, including personality, vocational interests, and societal attitudes. This is old news among the experts. The experts have, however, been startled by more recent discoveries.

One is that the heritability of IQ rises with age—that is to say, the extent to which genetics accounts for differences in IQ among individuals increases as people get older. Studies comparing identical and fraternal twins, published in the past decade by a group led by Thomas J. Bouchard, Jr., of the University of Minnesota and other scholars, show that about 40 percent of IQ differences among preschoolers stems from genetic differences but that heritability rises to 60 percent by adolescence and to 80 percent by late adulthood. With age, differences among individuals in their developed intelligence come to mirror more closely their genetic differences. It appears that the effects of environment on intelligence fade rather than grow with time. In hindsight, perhaps this should have come as no surprise. Young children have the circumstances of their lives imposed on them by parents, schools, and other agents of society, but as people get older they become more independent and tend to seek out the life niches that are most congenial to their genetic proclivities.

A second big surprise for intelligence experts was the discovery that environments shared by siblings have little to do with IQ. Many people still mistakenly believe that social, psychological, and economic differences among families create lasting and marked differences in IQ. Behavioral geneticists refer to such environmental effects as "shared" because they are common to siblings who grow up together. Research has shown that although shared environments do have a modest influence on IQ in childhood, their effects dissipate by adolescence. The IQs of adopted children, for example, lose all resemblance to those of their adoptive family members and become more like the IQs of the biological parents they have never known. Such findings suggest that siblings either do not share influential aspects of the rearing environment or do not experience them in the same way. Much behavioral genetics research currently focuses on the still mysterious processes by which environments make members of a household less alike.

■ *g* ON THE JOB

Although the evidence of genetic and physiological correlates of *g* argues powerfully for the existence of global intelligence, it has not quelled the critics of intelligence testing. These skeptics argue that even if such a global entity exists, it has no intrinsic functional value and becomes important only to the extent that people treat it as such: for example, by using IQ scores to sort, label, and assign students and employees. Such concerns over the proper use of mental tests have prompted a great deal of research in recent decades. This research shows that although IQ testes can indeed be misused, they measure a capability that does in fact affect many kinds of performance and many life outcomes, independent of the tests' interpretations or applications. Moreover, the research shows that intelligence tests measure the capability equally well for all native-born English-speaking groups in the U.S.

If we consider that intelligence manifests itself in everyday life as the ability to deal with complexity, then it is easy to see why it has great functional or practical importance. Children, for example, are regularly exposed to complex tasks once they begin school. Schooling requires above all that students learn, solve problems, and think abstractly. That IQ is quite a good predictor of differences in educational achievement is therefore not surprising. When scores on both IQ and standardized achievement tests in different subjects are averaged over several years, the two averages correlate as highly as different IQ tests for the same individual do. High-ability students also master material at many times the rate of their low-ability peers. Many investigators have helped quantify this discrepancy. For example, a 1969 study done for the U.S. Army by the Human Resources Research Office found that enlistees in the bottom fifth of the ability distribution required two to six times as many teaching trials and prompts as did their higher-ability peers to attain minimal proficiency in rifle assembly, monitoring signals, combat plotting, and other basic military tasks. Similarly, in school settings the ratio of learning rates between "fast" and "slow" students is typically five to one.

The scholarly content of many IQ tests and their strong correlations with educational success can give the impression that *g* is only a narrow academic ability. But general mental ability also predicts job performance, and in more complex jobs it does so better than any other single personal trait, including education and experi-

Life Chances	High Risk	Uphill Battle	Keeping Up	Out Ahead	Yours to Lose
Training Style	Slow, simple, supervised	Very explicit, hands-on	Written materials, plus experience	Gathers, infers own information	
			Mastery learning, hands-on	College format	
Career Potential		Assembler, food service, nurse's aide	Clerk, teller, police officer, machinist, sales	Manager, teacher, accountant	Attorney, chemist, executive
IQ	70	80 90	100 110	120	130

Population Percentages

Total population distribution	5	20	50	20	5
Out of labor force more than 1 month out of year (men)	22	19	15	14	10
Unemployed more than 1 month out of year (men)	12	10	7	7	2
Divorced in 5 years	21	22	23	15	9
Had illegitimate children (women)	32	17	8	4	2
Lives in poverty	30	16	6	3	2
Ever incarcerated (men)	7	7	3	1	0
Chronic welfare recipient (mothers)	31	17	8	2	0
High school dropout	55	35	6	0.4	0

Adapted from *Intelligence*, Vol.24, No.1:January/February 1997

Correlation of IQ scores with occupational achievement suggests that g reflects an ability to deal with cognitive complexity. Scores also correlate with some social outcomes (the percentages apply to young white adults in the U.S.). (John Mengel Ponzi & Weill)

ence. The army's Project A, a seven-year study conducted in the 1980s to improve the recruitment and training process, found that general mental ability correlated strongly with both technical proficiency and soldiering in the nine specialties studied, among them infantry, military police, and medical specialist. Research in the civilian sector has revealed the same pattern. Furthermore, although the addition of personality traits such as conscientiousness can help hone the prediction of job performance, the inclusion of specific mental aptitudes such as verbal fluency or mathematical skill rarely does. The predictive value of mental tests in the work arena stems almost entirely from their measurement of *g,* and that value rises with the complexity and prestige level of the job.

Half a century of military and civilian research has converged to draw a portrait of occupational opportunity along the IQ continuum. Individuals in the top 5 percent of the adult IQ distribution (above IQ 125) can essentially train themselves, and few occupations are beyond their reach mentally. Persons of average IQ (between 90 and 110) are not competitive for most professional and executive-level work but are easily trained for the bulk of jobs in the American economy. In contrast, adults in the bottom 5 percent of the IQ distribution (below 75) are very difficult to train and are not competitive for any occupation on the basis of ability. Serious problems in training low-IQ military recruits during World War II led Congress to ban enlistment from the lowest 10 percent (below 80) of the population, and no civilian occupation in modern economies routinely recruits its workers from that range. Current military enlistment standards exclude an individual whose IQ is below about 85.

The importance of *g* in job performance, as in schooling, is related to complexity. Occupations differ considerably in the complexity of their demands, and as that complexity rises, higher *g* levels become a bigger asset and lower *g* levels a bigger handicap. Similarly, everyday tasks and environments also differ significantly in their cognitive complexity. The degree to which a person's *g* level will come to bear on daily life depends on how much novelty and ambiguity that person's everyday tasks and surroundings present and how much continual leaning, judgment, and decision making they require. As gamblers, employers, and bankers know, even marginal differences in rates of return will yield big gains—or losses—over time. Hence, even small differences in *g* among people can exert large, cumulative influences across social and economic life.

In my own work, I have tried to synthesize the many lines of research that document the influence of IQ on life outcomes. As the illustration on page 65 shows, the odds of various kinds of achievement and social pathology change systematically across the IQ continuum, from borderline mentally retarded (below 70) to intellectually gifted (above 130). Even in comparisons of those of somewhat below average (between 76 and 90) and somewhat above average (between 111 and 125) IQs, the odds for outcomes having social consequence are stacked against the less able. Young men somewhat below average in general mental ability, for example, are more likely to be unemployed than men somewhat above average. The lower-IQ woman is four times more likely to bear illegitimate children than the higher-IQ woman; among mothers, she is eight times more likely to become a chronic welfare recipient. People somewhat below average are 88 times more likely to drop out of high school, seven times more likely to be jailed, and five times more likely as adults to live in poverty than people of somewhat above-average IQ. Below-average individuals are 50 percent more likely to be divorced than those in the above-average category.

These odds diverge even more sharply for people with bigger gaps in IQ, and the mechanisms by which IQ creates this divergence are not yet clearly understood. But no other single trait or circumstance yet studied is so deeply implicated in the nexus of bad social outcomes—poverty, welfare, illegitimacy, and educational failure—that entraps many low-IQ individuals and families. Even the effects of family background pale in comparison with the influence of IQ. As shown most recently by Charles Murray of the American Enterprise Institute in Washington, D.C., the divergence in many outcomes associated with IQ level is almost as wide among siblings from the same household as it is for strangers of comparable IQ levels. And siblings differ a lot in IQ—on average, by 12 points, compared with 17 for random strangers.

An IQ of 75 is perhaps the most important threshold in modern life. At that level, a person's chances of mastering the elementary school curriculum are only 50–50, and he or she will have a hard time functioning independently without considerable social support. Individuals and families who are only somewhat below average in IQ face risks of social pathology that, while lower, are still significant enough to jeopardize their well-being. High-IQ individuals may lack the resolve, character, or good fortune to capitalize on their intellectual capabilities, but socioeconomic success in the postindustrial information age is theirs to lose.

■ WHAT IS *VERSUS* WHAT COULD BE

The foregoing findings on g's effects have been drawn from studies conducted under a limited range of circumstances—namely, the social, economic, and political conditions prevailing now and in recent decades in developed countries that allow considerable personal freedom. It is not clear whether these findings apply to populations around the world, to the extremely advantaged and disadvantaged in the developing world, or, for that matter, to people living under restrictive political regimes. No one knows what research under different circumstances, in different eras, or with different populations might reveal.

But we do know that, wherever freedom and technology advance, life is an uphill battle for people who are below average in proficiency at learning, solving problems, and mastering complexity. We also know that the trajectories of mental development are not easily deflected. Individual IQ levels tend to remain unchanged from adolescence onward, and despite strenuous efforts over the past half a century, attempts to raise g permanently through adoption or educational means have failed. If there is a reliable, ethical way to raise or equalize levels of g, no one has found it.

Some investigators have suspected that biological interventions, such as dietary supplements of vitamins, may be more effective than educational ones in raising g levels. This approach is based in part on the assumption that improved nutrition has caused the puzzling rise in average levels of both IQ and height in the developed world during this century. Scientists are still hotly debating whether the gains in IQ actually reflect a rise in g or are caused instead by changes in less critical, specific mental skills. Whatever the truth may be, the differences in mental ability among individuals remain, and the conflict between equal opportunity and equal outcome persists. Only by accepting these hard truths about intelligence will society find humane solutions to the problems posed by the variations in general mental ability.

—*Scientific American Presents*, Winter 1998

THE GENETICS
OF COGNITIVE ABILITIES
AND DISABILITIES

Investigations of specific cognitive skills can help clarify
how genes shape the components of intellect

Robert Plomin and John C. DeFries

People differ greatly in all aspects of what is casually known as intelligence. The differences are apparent not only in school, from kindergarten to college, but also in the most ordinary circumstances: in the words people use and comprehend, in their differing abilities to read a map or follow directions, or in their capacities for remembering telephone numbers or figuring change. The variations in these specific skills are so common that they are often taken for granted. Yet what makes people so different?

It would be reasonable to think that the environment is the source of differences in cognitive skills—that we are what we learn. It is clear, for example, that human beings are not born with a full vocabulary; they have to learn words. Hence, learning must be the mechanism by which differences in vocabulary arise among individuals. And differences in experience—say, in the extent to which parents model and encourage vocabulary skills or in the quality of language training provided by schools—must be responsible for individual differences in learning.

Earlier in this century psychology was in fact dominated by environmental explanations for variance in cognitive abilities. More recently, however, most psychologists have begun to embrace a more balanced view: one in which nature and nurture interact in cognitive development. During the past few decades, studies in genetics have pointed to a substantial role for heredity in molding the components of intellect, and researchers have even begun to track down the genes involved in cognitive function. These findings do not refute the notion that environmental factors shape the learning process. Instead they suggest that differences in people's genes affect how easily they learn.

Just how much do genes and environment matter for specific cognitive abilities such as vocabulary? That is the question we have set out to answer. Our tool of study is quantitative genetics, a statistical approach that explores the causes of variations in traits among individuals. Studies comparing the performance of twins and adopted children on certain tests of cognitive skills, for example, can assess the relative contributions of nature and nurture.

In reviewing several decades of such studies and conducting our own, we have begun to clarify the relations among specialized aspects of intellect, such as verbal and spatial reasoning, as well as the relations between normal cognitive function and disabilities, such as dyslexia. With the help of molecular genetics, we and other investigators have also begun to identify the genes that affect these specific abilities and disabilities. Eventually, we believe, knowledge of these genes will help reveal the biochemical mechanisms involved in human intelligence. And with the insight gained from genetics, researchers may someday develop environmental interventions that will lessen or prevent the effects of cognitive disorders.

Some people find the idea of a genetic role in intelligence alarming or, at the very least, confusing. It is important to understand from the outset, then, what exactly geneticists mean when they talk about genetic influence. The term typically used is "heritability": a statistical measure of the genetic contribution to differences among individuals.

■ VERBAL AND SPATIAL ABILITIES

Heritability tells us what proportion of individual differences in a population—known as variance—can be ascribed to genes. If we say, for example, that a trait is

50 percent heritable, we are in effect saying that half of the variance in that trait is linked to heredity. Heritability, then, is a way of explaining what makes people different, not what constitutes a given individuals's intelligence. In general, however, if heritability for a trait is high, the influence of genes on the trait in individuals would be strong as well.

Attempts to estimate the heritability of specific cognitive abilities began with family studies. Analyses of similarities between parents and their children and between siblings have shown that cognitive abilities run in families. Results of a family study done on specific cognitive abilities, which was conducted in Hawaii in the 1970s, helped to quantify this resemblance.

The Hawaii Family Study of Cognition was a collaborative project between researchers at the University of Colorado at Boulder and the University of Hawaii and involved more than 1,000 families and sibling pairs. The study determined correlations (a statistical measure of resemblance) between relatives on tests of verbal and spatial ability. A correlation of 1.0 would mean that the scores of family members were identical; a correlation of zero would indicate that the scores were no more similar than those of two people picked at random. Because children on average share half their genes with each parent and with siblings, the highest correlation in test scores that could be expected on genetic grounds alone would be 0.5.

The Hawaii study showed that family members are in fact more alike than unrelated individuals on measures of specific cognitive skills. The actual correlation for both verbal and spatial tests were, on average, about 0.25. These correlations alone, however, do not disclose whether cognitive abilities run in families because of genetics or because of environmental effects. To explore this distinction, geneticists rely on two "experiments": twinning (an experiment of nature) and adoption (a social experiment).

Twin studies are the workhorse of behavioral genetics. They compare the resemblance of identical twins, who have the same genetic makeup, with the resemblance of fraternal twins, who share only about half their genes. If cognitive abilities are influenced by genes, identical twins ought to be more alike than fraternal twins on tests of cognitive skills. From correlations found in these kinds of studies, investigators can estimate the extent to which genes account for variances in the general population. Indeed, a rough estimate of heritability can be made by doubling the difference between identical-twin and fraternal-twin correlations.

Adoption provides the most direct way to disentangle nature and nurture in family resemblance, by creating pairs of genetically related individuals who do not share a common family environment. Correlations among these pairs enable investigators to estimate the contribution of genetics to family resemblance. Adoption also produces pairs of genetically unrelated individuals who share a family environ-

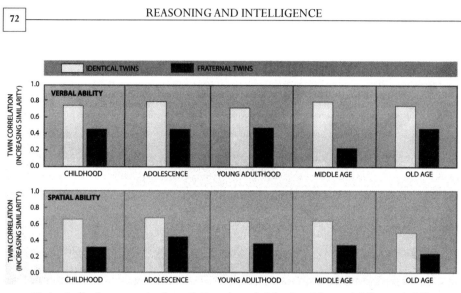

Twin studies have examined correlations in verbal (top) and in spatial (bottom) skills of identical twins and of fraternal twins. When the results of the separate studies are put side by side, they demonstrate a substantial genetic influence on specific cognitive abilities from childhood to old age; for all age groups, the scores of identical twins are more alike than those of fraternal twins. These data seem to counter the long-standing notion that the influence of genes wanes with time. (Jennifer C. Christiansen)

ment, and their correlations make it possible to estimate the contribution of shared environment to resemblance.

Twin studies of specific cognitive abilities over three decades and in four countries have yielded remarkably consistent results [*see illustration above*]. Correlations for identical twins greatly exceed those for fraternal twins on tests of both verbal and spatial abilities in children, adolescents, and adults. Results of the first twin study in the elderly, reported by Gerald F. McClearn and his colleagues at Pennsylvania State University and by Stig Berg and his associates at the Institute for Gerontology in Jönköping, Sweden, show that the resemblances between identical and fraternal twins persist even into old age. Although gerontologists have assumed that genetic differences become less important as experiences accumulate over a lifetime, research on cognitive abilities has so far demonstrated otherwise. Calculations based on the combined findings in these studies imply that in the general population genetics accounts for about 60 percent of the variance in verbal ability and about 50 percent of the variance in spatial ability.

Investigations involving adoptees have yielded similar results. Two recent studies of twins reared apart—one by Thomas J. Bouchard, Jr., Matthew McGue, and their colleagues at the University of Minnesota, the other an international collaboration headed by Nancy L. Pedersen at the Karolinska Institute in Stockholm—have implied heritabilities of about 50 percent for both verbal and spatial abilities.

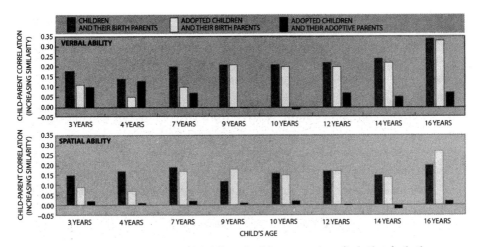

Colorado Adoption Project, which followed subjects over time, finds that for both ver-bal (top) and spatial (bottom) abilities, adopted children come to resemble their birth parents (white bars) as much as children raised by their birth parents do (gray bars). In contrast, adopted children do not end up resembling their adoptive parents (black bars). The results imply that most of the family resemblance in cognitive skills is caused by genetic factors, not environment. (Jennifer C. Christiansen)

In our own Colorado Adoption Project, which we launched in 1975, we have used the power of adoption studies to further characterize the roles of genes and environment, to assess developmental trends in cognitive abilities, and to explore the extent to which specific cognitive skills are related to one another. The ongoing project compares the correlations between more than 200 adopted children and their birth and adoptive parents with the correlations for a control group of children raised by their biological parents [*see illustration above*].

These data provide some surprising insights. By middle childhood, for example, birth mothers and their children who were adopted by others are just as similar as control parents and their children on measures of both verbal and spatial ability. In contrast, the scores of adopted children do not resemble those of their adoptive parents at all. These results join a growing body of evidence suggesting that the common family environment generally does not contribute to similarities in family members. Rather family resemblance on such measures seems to be controlled almost entirely by genetics, and environmental factors often end up making family members different, not the same.

The Colorado data also reveal an interesting developmental trend. It appears that genetic influence increases during childhood, so that by the mid-teens, heritability reaches a level comparable with that seen in adults. In correlations of verbal ability, for example, resemblance between birth parents and their children who were adopted by others increases from about 0.1 at age three to about 0.3 at age 16. A similar pattern is evident in tests of spatial ability. Some genetically driven transformation in cognitive function seems to take place in the early school years, around age seven. The results indicate that by the time people reach age 16, genetic factors account for 50 percent of the variance for verbal ability and 40 percent for spatial ability—numbers not unlike those derived from twin studies of specific cognitive abilities.

The Colorado Adoption Project and other investigations have also helped clarify the differences and similarities among cognitive abilities. Current cognitive neuroscience assumes a modular model of intelligence, in which different cognitive processes are isolated anatomically in discrete modules in the brain. The modular model implies that specific cognitive abilities are also genetically distinct—that genetic effects on verbal ability, say, should not overlap substantially with genetic effects on spatial ability.

Psychologists, however, have long recognized that most specialized cognitive skills, including verbal and spatial abilities, intercorrelate moderately. That is, people who perform well on one type of test also tend to do well on other types. Correlations between verbal and spatial abilities, for example, are usually about 0.5. Such intercorrelation implies a potential genetic link.

■ FROM ABILITIES TO ACHIEVEMENT

Genetic studies of specific cognitive abilities also fail to support the modular model. Instead it seems that genes are responsible for most of the overlap between cognitive skills. Analysis of the Colorado project data, for example, indicates that genetics governs 70 percent of the correlation between verbal and spatial ability. Similar results have been found in twin studies in childhood, young adulthood, and middle age. Thus, there is a good chance that when genes associated with a particular cognitive ability are identified, the same genes will be associated with other cognitive abilities.

Research into school achievement has hinted that the genes associated with cognitive abilities may also be relevant to academic performance. Studies of more than 2,000 pairs of high school–age twins were done in the 1970s by John C. Loehlin of the University of Texas at Austin and Robert C. Nichols, then at the National Merit Scholarship Corporation in Evanston, Illinois. In these studies the scores of identical twins were consistently and substantially more similar than those of fraternal twins on all four domains of the National Merit Scholarship Qualifying Test: English usage, mathematics, social studies, and natural sciences. These results suggest that genetic factors account for about 40 percent of the variation on such achievement tests.

Genetic influence on school achievement has also been found in twin studies of elementary school–age children as well as in our work with the Colorado Adoption Project. It appears that genes may have almost as much effect on school achievement as they do on cognitive abilities. These results are surprising in and of themselves, as educators have long believed that achievement is more a product of effort than of ability. Even more interesting, then, is the finding from twin studies and our adoption project that genetic effects overlap between different categories of achievement and that these overlapping genes are probably the very same genetic factors that can influence cognitive abilities.

This evidence supports a decidedly nonmodular view of intelligence as a pervasive or global quality of the mind and underscores the relevance of cognitive abilities in real-world performance. It also implies that genes for cognitive abilities are likely to be genes involved in school achievement, and vice versa.

Given the evidence for genetic influence on cognitive abilities and achievement, one might suppose that cognitive disabilities and poor academic achievement must also show genetic influence. But even if genes are involved in cognitive disorders, they may not be the same genes that influence normal cognitive function. The example of mental retardation illustrates this point. Mild mental retardation runs in families, but severe retardation does not. Instead severe mental retardation is caused by genetic and environmental factors—novel mutations, birth complications, and head injuries, to name a few—that do not come into play in the normal range of intelligence.

Researchers need to assess, rather than assume, genetic links between the normal and the abnormal, between the traits that are part of a continuum and true disorders of human cognition. Yet genetic studies of verbal and spatial disabilities have been few and far between.

■ GENETICS AND DISABILITY

Most such research has focused on reading disability, which afflicts 80 percent of children diagnosed with a learning disorder. Children with reading disability, also known as dyslexia, read slowly, show poor comprehension, and have trouble reading aloud. Studies by one of us (DeFries) have shown that reading disability runs in families and that genetic factors do indeed contribute to the resemblance among family members. The identical twin of a person diagnosed with reading disability, for example, has a 68 percent risk of being similarly diagnosed, whereas a fraternal twin has only a 38 percent chance.

Is this genetic effect related in any way to the genes associated with normal variation in reading ability? That question presents some methodological challenges. The concept of a cognitive disorder is inherently problematic, because it treats disability qualitatively—you either have it or you don't—rather than describing the degree of disability in a quantitative fashion. This focus creates an analytical gap between disorders and traits that are dimensional (varying along a continuum), which are by definition quantitative.

During the past decade, a new genetic technique has been developed that bridges the gap between dimensions and disorders by collecting quantitative information about the relatives of subjects diagnosed qualitatively with a disability. The method is called DF extremes analysis, after its creators, DeFries and David W. Fulker, a colleague at the University of Colorado's Institute for Behavioral Genetics.

For reading disability, the analysis works by testing identical and fraternal twins of reading-disabled subjects on quantitative measures of reading, rather than look-ing for a shared diagnosis of dyslexia [*see illustration below*]. If reading disability is in-fluenced by genes that also affect variation within the normal range of reading per-formance, then the reading score of the identical twins of dyslexic children should be closer to those of the reading-disabled group than the scores of fraternal twins are. (A single gene can exert different effects if it occurs in more than one form in a population, so that two people may inherit somewhat different versions. The genes controlling eye color and height are examples of such variable genes.)

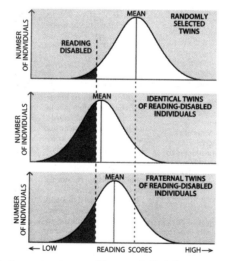

Reading scores of twins suggest a possible genetic link between normal and abnormal reading skills. In a group of randomly selected members of twin pairs (top), a small fraction of children were reading disabled. Identical (middle) and fraternal (bottom) twins of the reading-disabled children scored lower than the randomly selected group, with the identical twins performing worse than the fraternal ones. Genetic factors, then, are involved in reading disability. The same genes that influence reading disabil-ity may underlie differences in normal reading ability. (Jennifer C. Christiansen)

It turns out that, as a group, identical twins of reading-disabled subjects do perform almost as poorly as dyslexic subjects on these quantitative tests, whereas fraternal twins do much better than the reading-disabled group (though still significantly worse than the rest of the population). Hence the genes involved in reading disability may in fact be the same as those that contribute to the quantitative dimension of reading ability measured in this study. DF extremes analysis of these data further suggests that about half the difference in reading scores between dyslexics and the general population can be explained by genetics.

For reading disability, then, there could well be a genetic link between the normal and the abnormal, even though such links may not be found universally for other disabilities. It is possible that reading disability represents the extreme end of a continuum of reading ability, rather than a distinct disorder—that dyslexia might be quantitatively rather than qualitatively different from the normal range of reading ability. All this suggests that if a gene is found for reading disability, the same gene is likely to be associated with the normal range of variation in reading ability. The definitive test will come when a specific gene is identified that is associated with either reading ability or disability. In fact, we and other investigators are already very close to finding such a gene.

■ THE HUNT FOR GENES

Until now, we have confined our discussion to quantitative genetics, a discipline that measures the heritability of traits without regard to the kind and number of genes involved. For information about the genes themselves, researchers must turn to molecular genetics—and increasingly, they do. If scientists can identify the genes involved in behavior and characterize the proteins that the genes code for, new interventions for disabilities become possible.

Research in mice and fruit flies has succeeded in identifying single genes related to learning and spatial perception, and investigations of naturally occurring variations in human populations have found mutations in single genes that result in general mental retardation. These include the genes for phenylketonuria and fragile X syndrome, both causes of mental retardation. Single-gene defects that are associated with Duchenne's muscular dystrophy, Lesch-Nyhan syndrome, neurofibromatosis type 1 and Williams syndrome may also be linked to the specific cognitive disabilities seen in these disorders.

In fact, more than 100 single-gene mutations are known to impair cognitive development. Normal cognitive functioning, on the other hand, is almost certainly orchestrated by many subtly acting genes working together, rather than by single genes operating in isolation. These collaborative genes are thought to affect cognition in a probabilistic rather than a deterministic manner and are called quantitative trait loci, or QTLs. The name, which applies to genes involved in a complex dimension such as cognition, emphasizes the quantitative nature of certain physical and behavioral traits. QTLs have already been identified for diseases such as diabetes, obesity, and hypertension as well as for behavioral problems involving drug sensitivity and dependence.

But finding QTLs is much more difficult than identifying the single-gene mutations responsible for some cognitive disorders. Fulker addressed this problem by developing a method, similar to DF extremes analysis, in which certain known variations in DNA are correlated with sibling differences in quantitative traits. Because genetic effects are easier to detect at the extremes of a dimension, the method works best when at least one member of each sibling pair is known to be extreme for a trait. Investigators affiliated with the Colorado Learning Disabilities Research Center at the University of Colorado first used this technique, called QTL linkage, to try to locate a QTL for reading disability—and succeeded. The discovery was reported in 1994 by collaborators at Boulder, the University of Denver, and Boys Town National Research Hospital in Omaha.

Like many techniques in molecular genetics, QTL linkage works by identifying differences in DNA markers: stretches of DNA that are known to occupy particular sites on chromosomes and that can vary somewhat from person to person. The different versions of a marker, like the different versions of a gene, are called alleles. Because people have two copies of all chromosomes (except for the gender-determining X and Y chromosomes in males), they have two alleles for any given DNA marker. Hence, siblings can share one, two, or no alleles of a marker. In other words, for each marker, siblings can either be like identical twins (sharing both alleles), like fraternal twins (sharing half their alleles), or like adoptive siblings (sharing no alleles).

The investigators who found the QTL for reading disability identified a reading-disabled member of a twin pair and then obtained reading scores for the other twin—the "co-twin." If the reading scores of the co-twins were worse when they shared alleles of a particular marker with their reading-disabled twins, then that marker was likely to lie near a QTL for reading disability in the same chromosomal region. The researchers found such a marker on the short arm of chromosome 6 in two independent samples, one of fraternal twins and one of nontwin siblings. The findings have since been replicated by others.

It is important to note that whereas these studies have helped point to the location of a gene (or genes) implicated in reading disability, the gene (or genes) has not yet been characterized. This distinction gives a sense of where the genetics of cognition stand today: posed on the brink of a new level of discovery. The identification

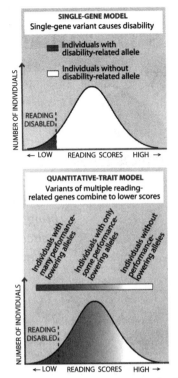

Two models illustrate how genetics may affect reading disability. In the classic view (top), a single variant, or allele of a gene is able to cause the disorder; everyone who has that allele becomes reading disabled (graph). But evidence points to a different model (bottom), in which a single allele cannot produce the disability on its own. Instead variants of multiple genes each act subtly but can combine to lower scores and increase the risk of disability. (Jennifer C. Christiansen)

of genes that influence specific cognitive abilities will revolutionize researchers' understanding of the mind. Indeed, molecular genetics will have far-ranging consequences for the study of all human behavior. Researchers will soon be able to investigate the genetic connections between different traits and between behaviors and biological mechanisms. They will be able to better track the developmental course of genetic effects and to define more precisely the interactions between genes and the environment.

The discovery of genes for disorders and disabilities will also help clinicians design more effective therapies and to identify people at risk long before the appearance of symptoms. In fact, this scenario is already being enacted with an allele called Apo-E4, which is associated with dementia and cognitive decline in the elderly. Of course, new knowledge of specific genes could turn up new problems as well: among them, prejudicial labeling and discrimination. And genetics research always raises fears that DNA markers will be used by parents prenatally to select "designer babies."

We cannot emphasize too much that genetic effects do not imply genetic determinism, nor do they constrain environmental interventions. Although some readers may find our views to be controversial, we believe the benefits of identifying genes for cognitive dimensions and disorders will far outweigh the potential abuses.

—May 1998

UNCOMMON TALENTS:
GIFTED CHILDREN,
PRODIGIES, AND SAVANTS

Possessing abilities well beyond their years,
gifted children inspire admiration, but they also
suffer ridicule, neglect, and misunderstanding

Ellen Winner

One evening a few years ago, while I was attending a concert, a young boy in the audience caught my attention. As the orchestra played a Mozart concerto, this nine-year-old child sat with a thick, well-thumbed orchestral score opened on his lap. As he read, he hummed the music out loud, in perfect tune. During intermission, I cornered the boy's father. Yes, he told me, Stephen was really reading the music, not just looking at it. And reading musical scores was one of his preferred activities, vying only with reading college-level computer programming manuals. At an age when most children concentrate on fourth-grade arithmetic and the nuances of playground etiquette, Stephen had already earned a prize in music theory that is coveted by adults.

Gifted children like Stephen are fascinating but also intimidating. They have been feared as "possessed," they have been derided as oddballs, they have been ridiculed as nerds. The parents of such young people are often criticized for pushing their children rather than allowing them a normal, well-balanced childhood. These children are so different from others that schools usually do not know how to educate them. Meanwhile society expects gifted children to become creative intellectuals and artists as adults and views them as failures if they do not.

Psychologists have always been interested in those who deviate from the norm, but just as they know more about psychopathology than about leadership and courage, researchers also know far more about retardation than about giftedness. Yet an understanding of the most talented minds will provide both the key to educating gifted children and a precious glimpse of how the human brain works.

■ THE NATURE OF GIFTEDNESS

Everyone knows children who are smart, hardworking achievers—youngsters in the top 10 to 15 percent of all students. But only the top 2 to 5 percent of children are gifted. Gifted children (or child prodigies who are just extreme versions of gifted children) differ from bright children in at least three ways:

- *Gifted children are precocious.* They master subjects earlier and learn more quickly than average children do.
- *Gifted children march to their own drummer.* They make discoveries on their own and can often intuit the solution of a problem without going through a series of logical, linear steps.
- *Gifted children are driven by "a rage to master."* They have a powerful interest in the area, or domain, in which they have high ability—mathematics, say, or art— and they can readily focus so intently on work in this domain that they lose sense of the outside world.

These are children who seem to teach themselves to read as toddlers, who breeze through college mathematics in middle school or who draw more skillfully as second-graders than most adults do. Their fortunate combination of obsessive interest and an ability to learn easily can lead to high achievement in their chosen domain. But gifted children are more susceptible to interfering social and emotional factors than once was thought.

The first comprehensive study of the gifted, carried out over a period of more than 70 years, was initiated at Stanford University in the early part of this century by Lewis M. Terman, a psychologist with a rather rosy opinion of gifted children. His study tracked more than 1,500 high-IQ children over the course of their lives. To qualify for the study, the "Termites" were first nominated by their teachers and then had to score 135 or higher on the Stanford-Binet IQ test (the average score is 100). These children were precocious: they typically spoke early, walked early, and read before they entered school. Their parents described them as being insatiably curious and as having superb memories.

Terman described his subjects glowingly, not only as superior in intelligence to other children but also as superior in health, social adjustment, and moral attitude. This conclusion easily gave rise to the myth that gifted children are happy and well adjusted by nature, requiring little in the way of special attention—a myth that still guides the way these children are educated today.

In retrospect, Terman's study was probably flawed. No child entered the study unless nominated by a teacher as one of the best and the brightest; teachers probably overlooked those gifted children who were misfits, loners, or problematic to teach. And the shining evaluations of social adjustment and personality in the gifted were performed by the same admiring teachers who had singled out the study subjects. Finally, almost a third of the sample came from professional, middle-class families. Thus, Terman confounded IQ with social class.

The myth of the well-adjusted, easy-to-teach gifted child persists despite more recent evidence to the contrary. Mihaly Csikszentmihalyi of the University of Chicago has shown that children with exceptionally high abilities in any area—not just in academics but in the visual arts, music, even athletics—are out of step with their peers socially. These children tend to be highly driven, independent in their thinking, and introverted. They spend more than the usual amount of time alone, and although they derive energy and pleasure from their solitary mental lives, they also report feeling lonely. The more extreme the level of gift, the more isolated these children feel.

Contemporary researchers have estimated that about 20 to 25 percent of profoundly gifted children have social and emotional problems, which is about twice the normal rate; in contrast, moderately gifted children do not exhibit a higher than average rate. By middle childhood, gifted children often try to hide their abilities in the hopes of becoming more popular. One group particularly at risk for such underachievement is academically gifted girls, who report more depression, lower self-esteem, and more psychosomatic symptoms than academically gifted boys do.

The combination of precocious knowledge, social isolation, and sheer boredom in many gifted children is a tough challenge for teachers who must educate them alongside their peers. Worse, certain gifted children can leap years ahead of their peers in one area yet fall behind in another. These children, the unevenly gifted, sometimes seem hopelessly out of sync.

■ THE UNEVENLY GIFTED

Terman was a proponent of the view that gifted children are globally gifted—evenly talented in all academic areas. Indeed, some special children have exceptional verbal skills as well as strong spatial, numerical, and logical skills that enable them to excel in mathematics. The occasional child who completes college as an early teen—or even as a preteen—is likely to be globally gifted. Such children are easy to spot: they are all-around high achievers. But many children exhibit gifts in one area of study and are unremarkable or even learning disabled in others. These may be creative children who are difficult in school and who are not immediately recognized as gifted.

Unevenness in gifted children is quite common. A recent survey of more than 1,000 highly academically gifted adolescents revealed that more than 95 percent show a strong disparity between mathematical and verbal interests. Extraordinarily strong mathematical and spatial abilities often accompany average or even deficient verbal abilities. Julian Stanley of Johns Hopkins University has found that many gifted children selected for special summer programs in advanced math have enormous discrepancies between their math and verbal skills. One such eight-year-old scored 760 out of a perfect score of 800 on the math part of the Scholastic Assessment Test (SAT) but only 280 out of 800 on the verbal part.

In a retrospective analysis of 20 world-class mathematicians, psychologist Benjamin S. Bloom, then at the University of Chicago, reported that none of his subjects had learned to read before attending school (yet most academically gifted children do read before school) and that six had had trouble learning to read. And a retrospective study of inventors (who presumably exhibit high mechanical and spatial aptitude) showed that as children these individuals struggled with reading and writing.

Indeed, many children who struggle with language may have strong spatial skills. Thomas Sowell of Stanford University, an economist by training, conducted a study of late-talking children after he raised a son who did not begin to speak until almost age four. These children tended to have high spatial abilities—they excelled at puzzles, for instance—and most had relatives working in professions that require strong spatial skills. Perhaps the most striking finding was that 60 percent of these children had engineers as first- or second-degree relatives.

The association between verbal deficits and spatial gifts seems particularly strong among visual artists. Beth Casey of Boston College and I have found that college art students make significantly more spelling errors than college students majoring either in math or in verbal areas such as English or history. On average, the art students not only misspelled more than half of a 20-word list but also made the kind of errors associated with poor reading skills—nonphonetic spellings such as "physicain" for "physician" (instead of the phonetic "fisician").

The many children who possess a gift in one area and are weak or learning disabled in others present a conundrum. If schools educate them as globally gifted, these students will continually encounter frustration in their weak areas; if they are held back because of their deficiencies, they will be bored and unhappy in their strong fields. Worst, the gifts that these children do possess may go unnoticed entirely when frustrated, unevenly gifted children wind up as misfits or troublemakers.

■ SAVANTS: UNEVEN IN THE EXTREME

The most extreme cases of spatial or mathematical gifts coexisting with verbal deficits are found in savants. Savants are retarded (with IQs between 40 and 70) and are either autistic or show autistic symptoms. "Ordinary" savants usually possess one skill at a normal level, in contrast to their otherwise severely limited abilities. But the rarer savants—fewer than 100 are known—display one or more skills equal to prodigy level.

Savants typically excel in visual art, music, or lightning-fast calculation. In their domain of expertise, they resemble child prodigies, exhibiting precocious skills, independent learning, and a rage to master. For instance, the drawing savant named Nadia sketched more realistically at ages three and four than any known child prodigy of the same age. In addition, savants will often surpass gifted children in the accuracy of their memories.

Savants are like extreme versions of unevenly gifted children. Just as gifted children often have mathematical or artistic genius and language-based learning disabilities, savants tend to exhibit a highly developed visual-spatial ability alongside severe deficits in language. One of the most promising biological explanations for this syndrome posits atypical brain organization, with deficits in the left hemisphere of the brain (which usually controls language) offset by strengths in the right hemisphere (which controls spatial and visual skills).

According to Darold A. Treffert, a psychiatrist now in private practice in Fond du Lac, Wisconsin, the fact that many savants were premature babies fits well with this notion of left-side brain damage and resultant right-side compensation. Late in pregnancy, the fetal brain undergoes a process called pruning, in which a large number of excess neurons die off. But the brains of babies born prematurely may not have been pruned yet; if such brains experience trauma to the left hemisphere near the time of birth, numerous uncommitted neurons elsewhere in the brain might remain to compensate for the loss, perhaps leading to a strong right-hemisphere ability.

Such trauma to a premature infant's brain could arise many ways—from conditions during pregnancy, from lack of oxygen during birth, from the administration of too much oxygen afterward. An excess of oxygen given to premature babies can cause blindness in addition to brain damage; many musical savants exhibit the triad of premature birth, blindness, and strong right-hemisphere skill.

Gifted children most likely possess atypical brain organization to some extent as well. When average students are tested to see which part of their brain controls their verbal skills, the answer is generally the left hemisphere only. But when mathematically talented children are tested the same way, both the left and right hemispheres are implicated in controlling language—the right side of their brains participates in tasks ordinarily reserved for the left. These children also tend not to be strongly right-handed, an indication that their left hemisphere is not clearly dominant.

The late neurologist Norman Geschwind of Harvard Medical School was intrigued by the fact that individuals with pronounced right-hemisphere gifts (that is, in math, music, art) are disproportionately nonright-handed (left-handed or ambidextrous) and have higher than average rates of left-hemisphere deficits such as delayed onset of speech, stuttering, or dyslexia. Geschwind and his colleague Albert Galaburda theorized that this association of gift with disorder, which they called the "pathology of superiority," results from the effect of the hormone testosterone on the developing fetal brain.

Geschwind and Galaburda noted that elevated testosterone can delay development of the left hemisphere of the fetal brain; this in turn might result in compensatory right-hemisphere growth. Such "testosterone poisoning" might also account for the larger number of males than females who exhibit mathematical and spatial gifts, nonright-handedness, and pathologies of language. The researchers also noted that gifted children tend to suffer more than the usual frequency of immune disorders such as allergies and asthma; excess testosterone can interfere with the development of the thymus gland, which plays a role in the development of the immune system.

Testosterone exposure remains a controversial explanation for uneven gifts, and to date only scant evidence from the study of brain tissue exists to support the theory of damage and compensation in savants. Nevertheless, it seems certain that gifts are hardwired in the infant brain, as savants and gifted children exhibit extremely high abilities from a very young age—before they have spent much time working at their gift.

■ EMPHASIZING GIFTS

Given that many profoundly gifted children are unevenly talented, socially isolated, and bored with school, what is the best way to educate them? Most gifted programs today tend to target children who have tested about 130 or so on standard IQ tests, pulling them out of their regular classes for a few hours each week of general instruction or interaction. Unfortunately, these programs fail the most talented students.

Generally, schools are focusing what few resources they have for gifted education on the moderately academically gifted. These children make up the bulk of current "pull-out" programs: bright students with strong but not extraordinary abilities, who do not face the challenges of precocity and isolation to the same degree as the profoundly gifted. These children—and indeed most children—would be better served if schools instead raised their standards across the board.

Other nations, including Japan and Hungary, set much higher academic expectations for their children than the U.S. does; their children, gifted or not, rise to the challenge by succeeding at higher levels. The needs of moderately gifted children could be met by simply teaching them a more demanding standard curriculum.

The use of IQ as a filter for gifted programs also tends to tip these programs toward the relatively abundant, moderately academically gifted while sometimes overlooking profoundly but unevenly gifted children. Many of those children do poorly on IQ tests, because their talent lies in either math or language, but not both. Students whose talent is musical, artistic, or athletic are regularly left out as well. It makes more sense to identify the gifted by examining past achievement in specific areas rather than relying on plain-vanilla IQ tests.

Schools should then place profoundly gifted children in advanced courses in their strong areas only. Subjects in which a student is not exceptional can continue to be taught to the student in the regular classroom. Options for advanced classes include arranging courses especially for the gifted, placing gifted students alongside older students within their schools, registering them in college courses, or enrolling them in accelerated summer programs that teach a year's worth of material in a few weeks.

Profoundly gifted children crave challenging work in their domain of expertise and the companionship of individuals with similar skills. Given the proper stimulation and opportunity, the extraordinary minds of these children will flourish.

—*Scientific American Presents*, Winter 1998

III

Memory and Learning

WORKING MEMORY AND THE MIND

Anatomic and physiological studies of monkeys are locating the neural machinery involved in forming and updating internal representations of the outside world. Such representations form a cornerstone of the rational mind

Patricia S. Goldman-Rakic

The seeming simplicity of everyday life belies the enormously complex ongoing operations of the mind. Even routine tasks such as carrying on a conversation or driving to work draw on a mixture of current sensory data and stored knowledge that has suddenly become relevant. The combination of moment-to-moment awareness and instant retrieval of archived information constitutes what is called the working memory, perhaps the most significant achievement of human mental evolution. It enables humans to plan for the future and to string together thoughts and ideas, which has prompted Marcel Just and Patricia Carpenter of Carnegie Mellon University to refer to working memory as "the blackboard of the mind."

Until recently, the fundamental processes involved in such higher mental functions defied description in the mechanistic terms of science. Indeed, for the greater part of this century, neurobiologists often denied that such functions were accessible to scientific analysis or declared that they belonged strictly to the domain of psychology and philosophy. Within the past two decades, however, neuroscientists have made great advances in understanding the relation between cognitive processes and the anatomic organization of the brain. As a consequence, even global mental attributes such as thought and intentionality can now be meaningfully studied in the laboratory.

The ultimate goal of that work is extraordinarily ambitious. Eventually researchers such as myself hope to be able to analyze higher mental functions in terms of the coordinated activation of neurons in various structures in the brain. It should also be possible to identify the cells that mediate the activity of those structures. Such research will help explain the origin of mind. It may also lead to more complete descriptions of baffling mental disorders such as schizophrenia.

For many years, insight into the operation of the brain was stymied by the misconception that memory is a single entity that could be traced to a single structure or location. Since the 1950s, neuroscientists have increasingly come to appreciate that memory consists of multiple components constructed around a distributed network of neurons. According to present thinking, a form of memory known as associative memory acquires facts and figures and holds them in long-term storage. That knowledge is of no use, however, unless it can be accessed and brought to mind in order to influence current behavior.

Working memory complements associative memory by providing for the short-term activation and storage of symbolic information, as well as by permitting the manipulation of that information. A simple activity involving working memory is the carry-over operation in mental arithmetic, which requires temporarily storing a string of numbers and holding the sum of one addition in mind while calculating the next. More complex examples include planning a chess move or constructing a sentence. Working memory in humans is considered fundamental to language comprehension, to learning, and to reason.

Numerous lines of evidence indicate that the operations of working memory are carried out in a part of the brain known as the prefrontal lobes of the cerebral cortex. (Cortex derives from the Latin word meaning bark; the cerebral cortex consists of an outer rind of so-called gray matter neurons surrounding the cerebrum.) Much of the evidence identifying this structure as the center for working memory comes from observations of the effects of injuries to the prefrontal part of the hemispheres. For example, patients having frontal lobe damage exhibit gross deficiencies in how they use knowledge to guide their behavior in everyday situations.

Nevertheless, they often retain a full store of information and may continue to score well on conventional tests of intelligence.

Although most fully developed in humans, some elements of working memory exist in other animals, especially in other primates; if their prefrontal cortices are damaged, those animals develop symptoms much like the ones seen in humans. Neuroscientists have therefore turned to monkeys in their efforts to explore the nature of working memory. Such exploration has been aided by the design of repeatable tests of working memory functions.

Working memory is being assessed in monkeys by means of tasks known as delayed-response tests, which evaluate an organism's ability to react to situations on the basis of stored or internalized representations rather than on information immediately present in the environment. In the prototypical delayed-response tests, an animal receives a brief visual or auditory stimulus that is then hidden or taken away. After a delay of several seconds, the animal is given a signal that tells it to respond to the location where the stimulus had appeared. If the response is correct, the animal receives a reward, usually food or juice.

Delayed-response tests tap working memory processes because the animal must retain the memory of the location of the stimulus during the period of the delay. The proper response at the end of the delay is indicated not by external stimuli but by the memory of what the subject saw on the previous trial. Furthermore, the correct response may differ from one trial to the next, depending on new information presented to the subject in each trial. Correct responses in working memory tasks, as in human affairs, are guided by memory rather than by immediate sensory information, and they depend on constant updating of the relevant information.

Delayed-response tests resemble very closely the object-performance task, developed in the early part of this century by the French child psychologist Jean Piaget, that is widely used to chart the cognitive development of young children. For Piaget's task, a child is shown two boxes, one of which contains a toy. The boxes are then closed. After a brief wait, during which the child is purposely distracted, the child is asked to pick which box contains the toy. Once the child gives several consecutive correct responses, the toy is switched into the other box while the child watches. The experimenter then continues the test to find out whether the child will change his or her response in accord with the updated information.

A series of studies has demonstrated that performance on the object-permanence task, like the ability to conduct delayed-response activities, depends on the degree of maturity of the subject's prefrontal cortex. Human infants less than about eight months old (whose cortices have not yet acquired adult circuitry) perform poorly on these tasks, as do monkeys whose prefrontal regions have been surgically ablated. In both cases, the subjects' responses are guided by habit and by

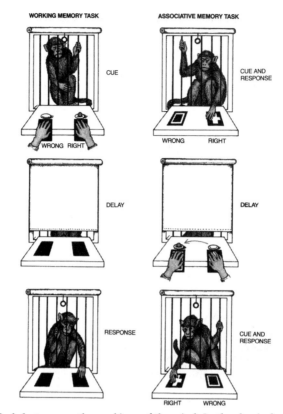

Memory tasks help to assess the workings of the mind. In the classical working memory task (left), a monkey briefly views a target stimulus—in this case, a morsel of food. Only after a delay is the animal allowed to retrieve the food. The experimenter randomly varies the location of the food between trials, so that each response tests only the animal's short-term retention of visual and spatial information. An associative memory test (right), in contrast, follows a consistent pattern throughout. Here a plus sign always indicates the correct response. The task therefore measures the animal's ability to retain long-term rules. (Patricia J. Wynne)

reflex rather than by representational principles. Infants and brain-injured monkeys tend to repeat the response that previously was reinforced (for example, choosing the box on the right even after they have seen that the toy was transferred to the box on the left) rather than change their response to agree with newly presented information. Both humans and monkeys act as if "out of sight" is "out of mind."

Such behavior implies that the mechanism for guiding behavior by representational knowledge is destroyed in monkeys having prefrontal lesions and not yet developed in human infants. In support of that notion, I, along with Jean-Pierre Bourgeois and Pasko Rakic, also at Yale University, have examined the rate at which neural connections form in the prefrontal cortices of juvenile monkeys.

The time of most rapid synapse formation in the animals' prefrontal region occurs when the animals are roughly two to four months old, the same age at which the monkeys acquire the capacity to perform delayed-response tasks. The concept that an object exists continuously in space and time even when out of view and, more generally, the ability to form abstract concepts may depend on a fundamental capacity to store representations of the outside world and to respond to those representations even when the real objects are not present.

The studies described above raised the enticing possibility of identifying more precisely the brain structures associated with delayed-response activities and representational memory. Much of the progress toward that goal has derived from experiments that monitor electrical activity in single neurons in monkeys' prefrontal cortices while the animals perform tasks that depend on specific delayed-response skills.

Joaquin M. Fuster of the University of California at Los Angeles, along with Kisou Kubota and Hiroaki Niki of the Kyoto Primate Center in Japan, performed the first experiments of how individual neurons behave in the prefrontal cortex. The researchers introduced fine electrodes into the prefrontal cortices of monkeys trained to perform simple delayed-response tasks and then recorded the animals' neuronal activity in relation to the events in the task. Those studies revealed a range of responses among the neurons in the prefrontal cortex. Some cells showed heightened electrical activity when information was presented, whereas others became active during the delay period, when the animals were remembering the information. A third set of neurons responded most strongly when the animals began their motor response.

At Yale, Shintaro Funahashi, Charles J. Bruce, and I have used the single-neuron technique in conjunction with a delayed-response experiment that tests spatial memory. For our experiment, a monkey is trained to fix its gaze on a small spot in the center of a television screen. A visual stimulus, typically a small square, appears briefly in one of eight locations on the screen and then vanishes. At the end of a delay of three to six seconds, the central light, or fixation spot, switches off, instructing the animal to move its eyes to the location where the stimulus was seen before the delay. If the response is correct, the animal is rewarded with a sip of grape juice. Because the animal's gaze is locked on to the fixation spot, each stimulus activates a specific set of retinal cells. Those cells, in turn, trigger only a certain subset of the visual pathways in the brain.

Using the eye-movement experiment, we have demonstrated that certain neurons in the prefrontal cortex possess what we call "memory fields": when a particular target disappears from view, an individual prefrontal neuron switches into an active state, producing electrical signals at more than twice the baseline rate. The neuron remains activated until the end of the delay period, when the animal deliv-

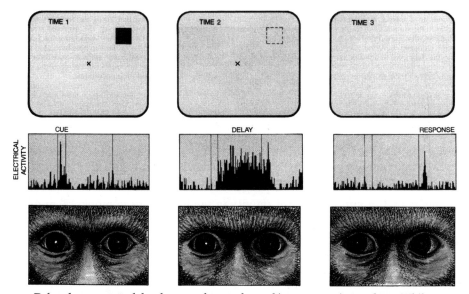

Delayed-response task has been used to study working memory in monkeys. While a monkey fixes its gaze on a central spot, a target flashes on the screen (left), then vanishes. During a delay of several seconds, the monkey keeps a memory of the spot "in mind" (center). When the central spot turns off, the animal moves its eyes to look where the target appeared (right). Measurements of electrical activity show that certain neurons in the prefrontal cortex react to the appearance of the target, others hold the memory of it in mind, and still others fire in preparation for a motor response. (Patricia J. Wynne)

ers its response. A given neuron appears always to code the same visual location. For example, some neurons fire only if the stimulus appears at the nine o'clock position on the television screen; the cell does not respond to visual stimuli that appear elsewhere in the monkey's visual field. Other neurons code for other target locations in working memory.

The neurons capable of retaining the visual and spatial coordinates of a stimulus (in other words, of keeping its location "in mind" after it vanishes) appear to be organized together within a specific area of the prefrontal cortex. These neurons collectively form the core of the spatial working memory system. If the activity of one or more of these neurons falters during the delay period—if the animal is distracted, for example—the animal will probably make an error.

The activation of prefrontal neurons during the delay period of a delayed-response task depends neither on the presence of an external stimulus nor on the execution of a response. Rather the neural activity corresponds to a mental event interposed between the stimulus and the response. Monkeys whose prefrontal cortices have been damaged have no difficulty in moving their eyes to a visible target or in reaching for a desired object, but they cannot direct those motor responses by remembering targets and objects that are no longer in evidence.

Because the prefrontal cortex functions as an intermediary between memory and action, one can imagine that damage to the prefrontal cortex could spare knowledge about the outside world yet destroy the organism's ability to bring that stored knowledge to mind and to utilize it. Indeed, monkeys whose prefrontal cortices have been damaged, as well as many humans with similar injuries, exhibit no difficulty learning sensory-discrimination tasks. All forms of associative, or long-term, learning are preserved as long as the subject can still find the familiar environmental stimuli associated with certain consequences and expectations.

Over the past decade, improved techniques for investigating the anatomy of the brain have provided for the first time an accurate and detailed picture of how the prefrontal cortex connects with major sensory and motor control centers. Various researchers have found that the part of the cortex near the principal sulcus, a large groove in the prefrontal cortex, is critical for the visual and spatial working memory functions. I have focused my research on this particular region in the belief that an in-depth neurobiological analysis of one major subdivision of the prefrontal cortex could serve as a starting point for analysis of the other subdivisions of the brain and help lead the way to development of a unified theory of the function of the entire prefrontal cortex.

Studies of direct and indirect neuronal linkages in the brain reveal that the prefrontal cortex is part of an elaborate network of reciprocal connections between the principal sulcus and the major sensory, limbic, and premotor areas of the cerebral cortex. That particular network seems to be dedicated to spatial information processing. The network's structure probably follows the same basic plan as do other similarly organized networks that draw on multiple parts of the brain and are dedicated to other cognitive functions—object recognition, learning production and comprehension, and mathematical reasoning, for example.

As previously noted, delayed-response experiments demonstrate that neurons in the principal sulcus are sensitive to the specific location of visual stimuli. Those neurons must therefore have access to visual and spatial information originating elsewhere in the brain. The principal sulcus does in fact receive signals from the posterior parietal cortex, where the brain processes spatial vision. Clinical studies have documented that damage to the parietal cortex in humans causes spatial neglect, a loss of awareness of the body and its relation to objects in the outside world.

Given that working memory depends on accessing and bringing to mind information that is stored in long-term memory, one might presume that the principal sulcus also interacts with the hippocampus, the neuronal structure that controls associative, or learned, memory. Researchers have used radioactive amino acids to trace direct connections between the principal sulcus and the hippocampus.

My colleague Harriet Friedman, also at Yale, and I have used a remarkable technique known as autoradiography to measure brain metabolism. Our work shows that the hippocampus and the principal sulcal areas of the cortex are often simultaneously active during delayed-response tasks. My co-workers and I think that the primary role of the hippocampus is to consolidate new associations, whereas the prefrontal cortex is necessary for retrieving the products of such associative learning (facts, events, rules) from long-term storage elsewhere in the brain for use in the task at hand.

A particularly useful version of autoradiography, called the 2-deoxyglucose method, has made it possible to observe directly which parts of the brain are activated during specific tasks. In this technique, developed by Louis Sokoloff of the National Institute of Mental Health, animals are injected with the compound 2-deoxyglucose, a molecule that appears chemically identical to glucose, the sugar that cells consume to provide energy. The more active a cell is, the more 2-deoxyglucose it takes in. Unlike normal glucose, however, 2-deoxyglucose cannot be broken down by metabolic activity, so it accumulates in the cell. Sokoloff uses a radioactive version of the compound. The concentration of radioactivity in each part of the brain is therefore directly proportional to how active the cells there have been.

For our studies, a monkey trained to perform the delayed-response tasks receives an intravenous injection of radioactive 2-deoxyglucose. Immediately after completing the task, the animal is sacrificed and its brain is dissected into thin slices that are placed on photographic film. Radioactivity darkens the film, so each exposure serves as a snapshot of the activity of the cells in one particular slice of the brain.

My colleagues and I have found that the prefrontal cortex, as well as many of the areas with which it is connected (for example, the hippocampus, the bottom portion of the parietal cortex, and the thalamus), exhibits a high level of metabolic activity during delayed-response performance. The same areas are notably less active when the monkey performs associative memory tasks that do not depend on short-term, rapid updating of information.

These results confirm anatomic studies of the connections between the prefrontal cortex and other parts of the brain. More significantly, they also reveal the degree to which various parts of the brain are engaged in certain discrete memory tasks. The studies also hint at how the prefrontal cortex organizes the many different kinds of information that must flood through it. In fact, patterns of brain activity appear distinctly different depending on whether the task calls up memories of location or of attributes of objects.

I think the prefrontal cortex is divided into multiple memory domains, each specialized for encoding a different kind of information, such as the location of objects, the features of objects (color, size, and shape), and additionally, in humans, semantic and mathematical knowledge. Recently Fraser Wilson and James Skelly in my laboratory at Yale have begun to define an area below the principal sulcus in monkeys where neurons respond preferentially to complex attributes of objects rather than to their locations. They have found neurons there that increase their rate of firing when a monkey is remembering a red circle but not when calling up a memory of a green square, for example.

Noninvasive imaging techniques are increasingly being used to monitor activation patterns in the human brain and to identify which neurons are engaged during specific mental tasks. One, known as positron emission tomography (PET), resembles autoradiography in that the subject takes in a radioactive compound that exposes changes in blood flow to a given region of the brain, indirectly displaying that region's degree of metabolic activity. Another way to record human brain activity is to measure the changing electrical potentials on the scalp in response to controlled sensory stimulation, a procedure called electroencephalography (EEG). Neither PET scans nor EEGs can provide anything close to the resolution possible in 2-deoxyglucose studies in animals, but they are invaluable tools for monitoring the human brain during mental activity.

A series of PET studies at Hammersmith Hospital in London and at Washington University examined subjects performing tasks that required them to keep a mental record of recently presented lists of words. Another PET experiment by the Washington University group required subjects to generate an appropriate verb to accompany a noun flashed in front of them on a card. The participants in all three tests displayed heightened neuronal activity in the prefrontal cortex while performing these tasks, all of which engaged working memory.

In a complementary study, Robert T. Knight of the University of California at Davis looked at EEGs of patients whose frontal lobes were injured. He asked them to perform tasks that depended on comparing current auditory stimuli with recently presented ones in order to detect whether they are the same or different. Frontal lobe patients displayed patterns of electrical activity quite unlike those of healthy subjects performing the same tasks, suggesting that the patients do not store recent information in memory the same way as do normal people.

In one study, subjects were exposed to steady patterns of low and high tones and occasional, unexpected auditory stimuli. Healthy people developed positive electrical potentials on their cortices within one third of a second of hearing the anomalous sound. Patients who had lesions in their prefrontal cortices showed no such response, although they reacted normally to the familiar background tones. These data are consistent with the notion that the prefrontal cortex temporarily stores information against which current stimuli are judged.

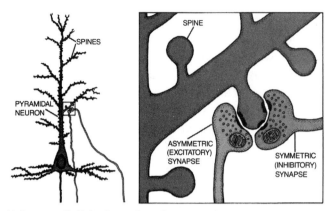

Pyramidal neuron (left) in the prefrontal cortex is thought to modulate signals to and from the prefrontal cortex. Each such neuron is covered with thousands of spines, bulb-like projections where synaptic connections occur. Synapses have different morphologies depending on whether they are excitatory or inhibitory (right). The dopamine-containing connections in the cortex are of the inhibitory type. (Patricia J. Wynne)

The ultimate function of the neurons in the prefrontal cortex is to excite or inhibit activity in other parts of the brain. In this way, information processed in the principal sulcus can direct neurons in the motor centers that in turn carry out movements of the eyes, mouth, hands, and other parts of the body. Whole-brain studies tell only part of the story; to understand the details of how signals pass to and from the prefrontal cortex, one must scrutinize the brain on a cellular scale.

When viewed through a conventional microscope, the cerebral cortex appears to be divided into six layers of varying cellular composition and density. Cells in each layer form their own set of connections within the brain. One class of cell, which resides in the fifth layer of the cortex, projects to areas beyond the cortex, including the caudate nucleus and putamen (which regulate a variety of motor activities) and the superior colliculus (which specifically processes visual motor functions). Neurons in the sixth layer of the cerebral cortex project into the thalamus, through which sensory inputs from the brain's periphery travel to reach the cortex.

The prefrontal cortex probably cannot independently trigger motor responses. Nevertheless, it may regulate motor behavior by initiating programming, facilitating, and canceling commands to brain structures that are more immediately involved in directing muscular movement. Such commands are transmitted via an elaborate set of chemical pathways in the brain. Neuroscientists and biochemists around the world have been racing to learn more about these chemicals and how they regulate the operation of the brain.

A number of researchers studying rodent brains, including Anne Marie Thierry and Jacques Glowinski of the College of France in Paris, Brigitte Berger of Pitié

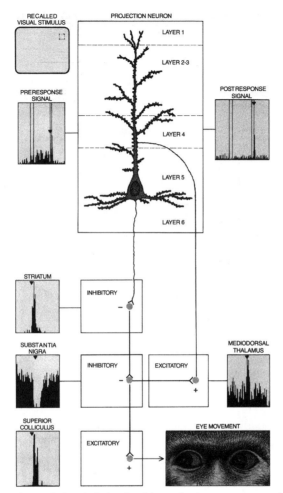

Elaborate flow of neural signals is involved in producing a memory-guided eye move-ment. A neuron in the fifth layer of the prefrontal cerebral cortex transmits signals along a chain of neurons in the striatum, the substantia nigra, and the superior col-liculus, where they trigger motor response in the eyes. Impulses from the substantia nigra travel to the mediodorsal thalamus and back to the cortex, indicating the comple-tion of the motor response and signaling the prefrontal neuron to return to a baseline level of activity. The graphs show the electrical activity of the neurons; inverted trian-gles indicate the nearly instantaneous travel of the signals. (Patricia J. Wynne)

Salpêtrière Hospital, also in Paris, and Tomas Hökfelt of the Karolinska Institute in Sweden, along with many colleagues, find that the prefrontal cortex abounds in catecholamines, a family of compounds that prepare the body for a stressful situa-tion. Those compounds also act as neurotransmitters, substances that transmit neu-ronal impulses in the brain. My co-workers and I have discovered a similar abundance of catecholamines in the prefrontal cortices of nonhuman primates. One of the most familiar catecholamines, dopamine, regulates how neurons react to stimuli and seems to play a central role in schizophrenia.

A growing body of evidence suggests that dopamine is one of the most important of the chemicals that regulate cell activity associated with working memory. An imbalance in the abundance of dopamine in the prefrontal cortex can induce deficits in the working memory similar to those resulting from lesions in the principal sulcus region of the prefrontal cortex. For example, aged monkeys whose prefrontal cortices are deficient in dopamine and norepinephrine (a chemical relative of adrenaline) perform poorly in delayed-response tests. Injecting the aged animals with the deficient neurotransmitters restored their memory function so that they tested roughly as well as younger, healthy monkeys.

Many of my colleagues and I are striving to learn which cells respond to dopamine and how they affect working memory. Within the past several years, we have collected evidence showing that neurons in certain layers of the cerebral cortex contain a great abundance of D_1 receptors, one of the chemical sites where dopamine binds to a cell. Interestingly, the neurons that are rich in D_1 receptors are those that project to the thalamus, the brain structure that relays information to the cortex.

Csaba Leranth, John Smiley, and F. Mark Williams of Yale are examining the cellular structures that enable dopamine to modulate responses to sensory inputs in the cerebral cortex. The researchers use an antibody developed by Michel Geffard of the Institute of Cellular Biochemistry and of Neurochemistry of the National Center of Scientific Research in Bordeaux, France, to label the neurons and their axonal projections that contain dopamine. They then scrutinize those cells under an electron microscope. The team looked in particular at the points of contact between dopamine-releasing cells and the neuronal spines, small protuberances where the cells receive incoming signals. Spines are discrete sites where calcium ions can enter and activate cellular mechanisms involving information processing and modulation of neuronal responses.

In most cases, the dopamine-releasing cells make symmetric contact with the spines—that is, the cell projections on either side of the synaptic cell show roughly the same density. Such symmetric contacts are thought to have an inhibitory effect: when the postsynaptic site is activated, the cells' normal, spontaneous electrical activity is dampened. A large proportion of the spines of pyramidal cells—the major class of neuron that projects out of the cortex—receive asymmetric contacts from the axons of another cell whose point of origin has not yet been identified but which is thought to carry signals from other cortical areas. Those asymmetric contacts probably have an opposite, excitatory effect.

Pyramidal cells receive the major sensory or informational signals arriving at the cerebral cortex. The network of excitatory and inhibitory synapses, or connections, noted by the Yale group provides a mechanism by which dopamine could alter the way that various classes of pyramidal neurons respond to integrate such signals across thousands of spines in their dendrites. In this way, dopamine may regulate the overall output of the cortex. Further analysis of the physical and chemical interactions between pyramidal cells and other neurons in the cerebral cortex should clarify how dopamine and other neurotransmitters influence cognition by stimulating or repressing the cellular responses of cortical neurons.

Investigations of the workings of the prefrontal cortex are revealing not only how the mind operates but also what goes wrong when it malfunctions. Medical researchers have implicated dysfunction of the prefrontal cortex as the cause of many neurological and psychiatric disorders, including Parkinson's disease and especially schizophrenia. The abnormal mental attributes associated with schizophrenia strongly resemble those caused by physical damage to the prefrontal cortex: thought disorders, reduced attention span, inappropriate or flattened emotional responses, and lack of initiative, plans, and goals. Schizophrenic patients, like frontal lobe patients and monkeys afflicted with prefrontal lobe lesions, retain a normal ability to perform routine procedures or habits but exhibit fragmented, disorganized behavior when attempting to perform tasks involving symbolic or verbal information.

Schizophrenic patients taking tests such as the Wisconsin Card Sort test tend to repeat a previous response even when it is clear that it is no longer the correct one; normal subjects, in contrast, shift hypotheses much sooner after making an error. Schizophrenic individuals are also severely impaired both on spatial delayed-response tasks and on a variety of tests of problem solving, abstraction, and planning.

Studies of cerebral blood flow by David H. Ingvar of University Hospital in Lund, Sweden, and by Daniel R. Weinberger, Karen F. Berman, and others at the National Institute of Mental Health, as well as measurements of local cerebral metabolism made by Monte S. Buchsbaum of the University of California at Irvine, show that schizophrenic patients have below-average blood flow into their prefrontal cortices, indicative of a depressed level of activity in that part of the brain. Schizophrenic subjects often suffer from impaired ability to move their eyes to track and project the forward trajectories of moving targets, further evidence that the disorder involves malfunctions in a posterior part of the prefrontal cortex, where the eye-movement centers involved to predictive tracking are located.

Sohee Park and Philip S. Holzman of Harvard University have shown that schizophrenic subjects exhibit impaired performance on working memory tasks much like those my colleagues and I have used to study working memory in rhesus monkeys. Conversely, Martha MacAvoy and Bruce of Yale have demonstrated that monkeys with lesions in the relevant portions of the prefrontal cortex exhibit the same type of predictive tracking disorder that has long been considered a marker of schizophrenia in humans.

Perhaps researchers should begin to think of schizophrenia as a breakdown in the process by which representational knowledge governs behavior. In my view, neural pathways in the prefrontal cortex update inner models of reality to reflect changing environmental demands and incoming information. Those pathways guide short-term memory and moment-to-moment behavior. If they fail, the brain views the world as a series of disconnected events, like a slide show, rather than as a continuous sequence, like a movie. The result is schizophrenic behavior, excessively dominated by immediate stimulation rather than by a balance of current, internal, and past information.

At present, theories describing the fundamental causes of schizophrenia are inadequate, much as knowledge of the functioning of the working memory system remains frustratingly sketchy. Fortunately, neurobiological research has been advancing at a breathless pace in the past few years. Such research should lead to a greater understanding not only of schizophrenia but of the prefrontal cortex and how it shapes short-term memory and the broader working of the rational mind.

—September 1992

EMOTION, MEMORY, AND THE BRAIN

*The neural routes underlying the formation
of memories about primitive emotional experiences,
such as fear, have been traced*

Joseph E. LeDoux

Despite millennia of preoccupation with every facet of human emotion, we are still far from explaining in a rigorous physiological sense this part of our mental experience. Neuroscientists have, in modern times, been especially concerned with the neural basis of cognitive processes such as perception and memory. They have for the most part ignored the brain's role in emotion.

Yet in recent years, interest in this mysterious mental terrain has surged. Catalyzed by breakthroughs in understanding the neural basis of cognition and by an increasingly sophisticated knowledge of the anatomical organization and physiology of the brain, investigators have begun to tackle the problem of emotion. One quite rewarding area of research has been the inquiry into the relation between memory and emotion. Much of this examination has involved studies of one particular emotion—fear—and the manner in which specific events or stimuli come, through individual learning experiences, to evoke this state. Scientists, myself included, have been able to determine the way in which the brain shapes how we form memories about this basic, but significant, emotional event. We call this process "emotional memory."

By uncovering the neural pathways through which a situation causes a creature to learn about fear, we hope to elucidate the general mechanisms of this form of memory. Because many human mental disorders—including anxiety, phobia, posttraumatic stress syndrome, and panic attack—involve malfunctions in the brain's ability to control fear, studies of the neural basis of this emotion may help us further understand and treat these disturbances.

Most of our knowledge about how the brain links memory and emotion has been gleaned through the study of so-called classical fear conditioning. In this process the subject, usually a rat, hears a noise or sees a flashing light that is paired with a brief, mild electric shock to its feet. After a few such experiences, the rat responds automatically to the sound or light even in the absence of the shock. Its reactions are typical to any threatening situation: the animal freezes, its blood pressure and heart rate increase, and it startles easily. In the language of such experiments, the noise or flash is a conditioned stimulus, the foot shock is an unconditioned stimulus, and the rat's reaction is a conditioned response, which consists of readily measured behavioral and physiological changes.

Conditioning of this kind happens quickly in rats—indeed, it takes place as rapidly as it does in humans. A single pairing of the shock to the sound or sight can bring on the conditioned effect. Once established, the fearful reaction is relatively permanent. If the noise or light is administered many times without an accompanying electric shock, the rat's response diminishes. This change is called extinction. But considerable evidence suggests that this behavioral alteration is the result of the brain's controlling the fear response rather than the elimination of the emotional memory. For example, an apparently extinguished fear response can recover spontaneously or can be reinstated by an irrelevant stressful experience. Similarly, stress can cause the reappearance of phobias in people who have been successfully treated. This resurrection demonstrates that the emotional memory underlying the phobia was rendered dormant rather than erased by treatment.

Fear conditioning has proved an ideal starting point for studies of emotional memory for several reasons. First, it occurs in nearly every animal group in which it has been examined: fruit flies, snails, birds, lizards, fish, rabbits, rats, monkeys, and people. Although no one claims that the mechanisms are precisely the same in all these creatures, it seems clear from studies to date that the pathways are very similar in mammals and possibly in all vertebrates. We therefore are confident in believing that many of the findings in animals apply to humans. In addition, the kinds of stimuli most commonly used in this type of conditioning are not signals that rats—or humans, for that matter—encounter in their daily lives. The novelty and irrelevance of these lights and sounds help to ensure that the animals have not already developed strong emotional reactions to them. So researchers are clearly observing learning and memory at work. At the same time, such cues do not require complicated cognitive processing from the brain. Consequently, the stimuli permit us to study emotional mechanisms relatively directly. Finally, our extensive knowledge of the neural pathways involved in processing acoustic and visual information serves as an excellent starting point for examining the neurological foundations of fear elicited by such stimuli.

My work has focused on the cerebral roots of learning fear, specifically fear that has been induced in the rat by associating sounds with foot shock. As do most other investigators in the field, I assume that fear conditioning occurs because the shock modifies the way in which neurons in certain important regions of the brain interpret the sound stimulus. These critical neurons are thought to be located in the neural pathway through which the sound elicits the conditioned response.

During the past 10 years, researchers in my laboratory, as well as in others, have identified major components of this system. Our study began when my colleagues at Cornell University Medical College, where I worked several years ago, and I asked a simple question: Is the auditory cortex required for auditory fear conditioning? In the auditory pathway, as in other sensory systems, the cortex is the highest level of processing; it is the culmination of a sequence of neural steps that starts with the peripheral sensory receptors located, in this case, in the ear. If lesions in, or surgical removal of, parts of the auditory cortex interfered with fear conditioning, we could conclude that the region is indeed necessary for this activity. We could also deduce that the next step in the conditioning pathway would be an output from the auditory cortex. But our lesion experiments confirmed what a series of other studies had already suggested: the auditory cortex is not needed in order to learn many things about simple acoustic stimuli.

We then went on to make lesions in the auditory thalamus and the auditory midbrain, sites lying immediately below the auditory cortex. Both these areas process auditory signals: the midbrain provides the major input to the thalamus; the thalamus supplies the major input to the cortex. Lesions in both regions completely eliminated the rat's susceptibility to conditioning. This discovery suggested that a sound stimulus is transmitted through the auditory system to the level of the auditory thalamus but that it does not have to reach the cortex in order for fear conditioning to occur.

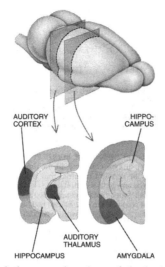

AUDITORY CORTEX

HIPPO-CAMPUS

AUDITORY THALAMUS

HIPPOCAMPUS

AMYGDALA

Anatomy of emotion includes several regions of the brain. Shown here in the rat (above), the amygdala, the thalamus, and parts of the cortex interact to create memories about fearful experiences associated, in this case, with sound. Recent work has located precise areas where fear is learned and remembered: certain parts of the thalamus communicate with areas in the amygdala that process the fear-causing sound stimuli. Because these neural mechanisms are thought to be similar in humans, the study of emotional memory in rodents may illuminate aspects of fear disorders in people. (Robert Osti)

This possibility was somewhat puzzling. We knew that the primary nerve fibers that carry signals from the auditory thalamus extend to the auditory cortex. So David A. Ruggiero, Donald J. Reis, and I looked again and found that, in fact, cells in some regions of the auditory thalamus also give rise to fibers that reach several subcortical locations. Could these neural projections be the connections through which the stimulus elicits the response we identify with fear? We tested this hypothesis by making lesions in each one of the subcortical regions with which these fibers connect. The damage had an effect in only one area: the amygdala.

That observation suddenly created a place for our findings in an already accepted picture of emotional processing. For a long time, the amygdala has been considered an important brain region in various forms of emotional behavior. In 1979 Bruce S. Kapp and his colleagues at the University of Vermont reported that lesions in the amygdala's central nucleus interfered with a rabbit's conditioned heart rate response once the animal had been given a shock paired with a sound. The central nucleus connects with areas in the brain stem involved in the control of heart rate, respiration, and vasodilation. Kapp's work suggested that the central nucleus was a crucial part of the system through which autonomic conditioned responses are expressed.

In a similar vein, we found that lesions of this nucleus prevented a rat's blood pressure from rising and limited its ability to freeze in the presence of a fear-causing stimulus. We also demonstrated, in turn, that lesions of areas to which the central nucleus connects eliminated one or the other of the two responses. Michael Davis and his associates at Yale University determined that lesions of the central nucleus, as well as lesions of another brain stem area to which the central nucleus project, diminished yet another conditioned response: the increased startle reaction that occurs when an animal is afraid.

The findings from various laboratories studying different species and measuring fear in different ways all implicated the central nucleus as a pivotal component of fear-conditioning circuitry. It provides connections to the various brain stem areas involved in the control of a spectrum of responses.

Despite our deeper understanding of this site in the amygdala, many details of the pathway remained hidden. Does sound, for example, reach the central nucleus directly from the auditory thalamus? We found that it does not. The central nucleus receives projections from thalamic areas next to, but not in, the auditory part of the thalamus. Indeed, an entirely different area of the amygdala, the lateral nucleus, receives inputs from the auditory thalamus. Lesions of the lateral nucleus prevented fear conditioning. Because this site gets information directly from the sensory system, we have come to think of it as the sensory interface of the amygdala in fear conditioning. In contrast, the central nucleus appears to be the interface with the systems that control responses.

These findings seemed to place us on the threshold of being able to map the entire stimulus response pathway. But we still did not know how information received by the lateral nucleus arrived at the central nucleus. Earlier studies had suggested that the lateral nucleus projects directly to the central nucleus, but the connections were fairly sparse. Working with monkeys, David Amaral and Asla Pitkanen of the Salk Institute for Biological Studies in San Diego demonstrated that the lateral nucleus extends directly to an adjacent site, called the basal or basolateral nucleus, which, in turn, projects to the central nucleus.

Collaborating with Lisa Stefanacci and other members of the Salk team, Claudia R. Farb and C. Genevieve Go in my laboratory at New York University found the same connections in the rat. We then showed that these connections form synaptic contacts—in other words, they communicate directly, neuron to neuron. Such contacts indicate that information reaching the lateral nucleus can influence the central nucleus via the basolateral nucleus. The lateral nucleus can also influence the central nucleus by way of the accessory basal or basomedial nucleus. Clearly, ample opportunities exist for the lateral nucleus to communicate with the central nucleus once a stimulus has been received.

The emotional significance of such a stimulus is determined not only by the sound itself but by the environment in which it occurs. Rats must therefore learn not only that a sound or visual cue is dangerous, but under what conditions it is so. Russell G. Phillips and I examined the response of rats to the chamber, or context, in which they had been conditioned. We found that lesions of the amygdala interfered with the animals' response to both the tone and the chamber. But lesions of the hippocampus—a region of the brain involved in declarative memory—interfered only with response to the chamber, not the tone. (Declarative memory involves explicit, consciously accessible information, as well as spatial memory.) At about the same time, Michael S. Fanselow and Jeansok J. Kim of the University of California at Los Angeles discovered that hippocampal lesions made after fear conditioning had taken place also prevented the expression of responses to the surroundings.

These findings were consistent with the generally accepted view that the hippocampus plays an important role in processing complex information, such as details about the spatial environment where activity is taking place. Phillips and I also demonstrated that the subiculum, a region of the hippocampus that projects to other areas of the brain, communicated with the lateral nucleus of the amygdala. This connection suggests that contextual information may acquire emotional significance in the same way that other events do—via transmission to the lateral nucleus.

Although our experiments had identified a subcortical sensory pathway that gave rise to fear conditioning, we did not dismiss the importance of the cortex. The interaction of subcortical and cortical mechanisms in emotion remains a hotly debated topic. Some researchers believe cognition is a vital precursor to emotional experience; others think that cognition—which is presumably a cortical function—is necessary to initiate emotion or that emotional processing is a type of cognitive processing. Still others question whether cognition is necessary for emotional processing.

It became apparent to us that the auditory cortex is involved in, though not crucial to, establishing the fear response, at least when simple auditory stimuli are applied. Norman M. Weinberger and his colleagues at the University of California at Irvine have performed elegant studies showing that neurons in the auditory cortex undergo specific physiological changes in their reaction to sounds as a result of conditioning. This finding indicates that the cortex is establishing its own record of the event.

Experiments by Lizabeth M. Romanski in my laboratory have determined that in the absence of the auditory cortex, rats can learn to respond fearfully to a single tone. If, however, projections from the thalamus to the amygdala are removed, projections from the thalamus to the cortex and then to the amygdala are sufficient. Romanski went on to establish that the lateral nucleus can receive input from both

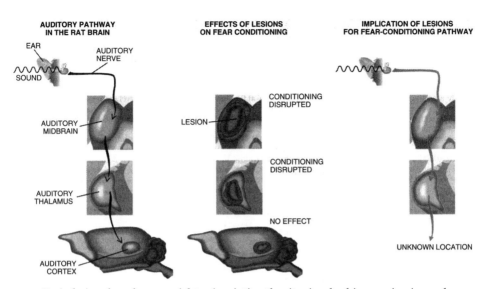

Brain lesions have been crucial to pinpointing the sites involved in experiencing and learning about fear. When a sound is processed by the rat brain, it follows a pathway from ear to midbrain to thalamus to cortex (left). Lesions can be made in various sites in the auditory pathway to determine which areas are necessary for fear conditioning (center). Only damage to the cortex does not disrupt the fear response, which suggests that some other areas of the brain receive the output of the thalamus and are involved in establishing memories about experiences that stimulate fear (right). (Ian Worpole)

the thalamus and the cortex. Her anatomical work in the rat complements earlier research in primates.

Theodore W. Jarrell and other workers in Neil Schneiderman's laboratory at the University of Miami have shown that lesions in the auditory cortex disrupt fear conditioning to one of two stimuli that was paired with foot shock. Rabbits expressed fear responses only to the sound that had been coupled with the shock. After receiving auditory cortex lesions, however, the animals responded to both tones. When the auditory cortex was absent and animals had to rely solely on the thalamus and the amygdala for learning, the two stimuli were indistinguishable. This work suggests that the cortex is not needed to establish simple fear conditioning; instead it serves to interpret stimuli when they become more intricate. Schneiderman's findings are supported by research in primates showing that projections to the amygdala from sensory regions of the cortex are important in processing the emotional significance of complex stimuli.

Some of this work has been challenged by the intriguing studies of Davis and his team. They reported that damage to a region of the perirhinal cortex—a transitional region between the older and newer cortex—prevents the expression of a previously learned fear response. Davis argues, therefore, that the cortex is the preferred pathway to the amygdala and that thalamic projections are not normally used during learning, unless the cortex is damaged at the time of learning. Our general understanding of the effect of lesions administered after learning has taken place is that they interfere with long-term memory storage or retrieval. This interpretation seems applicable to Davis's work as well and is suggested by recent studies by Keith P. Corodimas in my laboratory. He showed that at least part of the deficit can be eliminated by providing reminder cues.

Once we had a clear understanding of the mechanism through which fear conditioning is learned, we attempted to find out how emotional memories are established and stored on a molecular level. Farb and I showed that the excitatory amino acid transmitter glutamate is present in the thalamic cells that reach the lateral nucleus. Together with Chiye J. Aoki, we showed that it is also present at synapses in the lateral nucleus. Because glutamate transmission is implicated in memory formation, we seemed to be on the right track.

Glutamate has been observed in a process called long-term potentiation, or LTP, that has emerged as a model for the creation of memories. This process, which is most frequently studied in the hippocampus, involves a change in the efficiency of synaptic transmission along a neural pathway—in other words, signals travel more readily along this pathway once LTP has taken place. The mechanism seems to involve glutamate transmission and a class of postsynaptic excitatory amino acid receptors known as NMDA receptors.

Various studies have found LTP in the fear-conditioning pathway. Marie-Christine Clugnet and I noted that LTP could be included in the thalamo-amygdala pathway. Thomas H. Brown and Paul Chapman and their colleagues at Yale discovered LTP in a cortical projection to the amygdala. Other researchers, including Davis and Fanselow, have been able to block fear conditioning by clocking NMDA receptors in the amygdala. And Michael T. Rogan in my laboratory found that the processing of sounds by the thalamo-amygdala pathway is amplified after LTP has been induced. The fact that LTP can be demonstrated in a conditioning pathway offers new hope for understanding how LTP might relate to emotional memory.

In addition, recent studies by Fabio Bordi, also in my laboratory, have suggested hypotheses about what could be going on in the neurons of the lateral nucleus during learning. Bordi monitored the electrical state of individual neurons in this area when a rat was listening to the sound and receiving the shock. He and Romanski found that essentially every cell responding to the auditory stimuli also responded to the shock. The basic ingredient of conditioning is thus present in the lateral nucleus.

Bordi was able to divide the acoustically stimulated cells into two classes: habituating and consistently responsive. Habituating cells eventually stopped responding

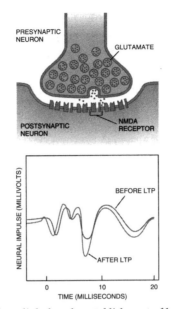

PRESYNAPTIC NEURON

GLUTAMATE

POSTSYNAPTIC NEURON

NMDA RECEPTOR

NEURAL IMPULSE (MILLIVOLTS)

BEFORE LTP

AFTER LTP

0   10   20
TIME (MILLISECONDS)

Memory formation has been linked to the establishment of long-term potentiation, or LTP. In this model of memory the neurotransmitter glutamate and its receptors, called NMDA receptors (top), bring about strengthened neural transmission. Once LTP is established, the same neural signals produce larger responses (bottom). Emotional memories may also involve LTP in the amygdala. Glutamate and NMDA receptors have been found in the region of the amygdala where fear conditioning takes place. (Ian Worpole)

to the repeated sound, suggesting that they might serve to detect any sound that was unusual or different. They could permit the amygdala to ignore a stimulus once it became familiar. Sound and shock pairing at these cells might reduce habituation, thereby allowing the cells to respond to, rather than ignore, significant stimuli.

The consistently responsive cells had high-intensity thresholds: only loud sounds could activate them. That finding is interesting because of the role loudness plays in judging distance. Nearby sources of sound are presumably more dangerous than those that are far away. Sound coupled with shock might act on these cells to lower their threshold, increasing the cells' sensitivity to the same stimulus. Consistently responsive cells were also broadly tuned. The joining of a sound and a shock could make the cells responsive to a narrower range of frequencies, or it could shift the tuning toward the frequency of the stimulus. In fact, Weinberger has shown that cells in the auditory system do alter their tuning to approximate the conditioned stimulus. Bordi and I have detected this effect in lateral nucleus cells as well.

The apparent permanence of these memories raises an important clinical question: Can emotional learning be eliminated, and, if not, how can it be toned down? As noted earlier, it is actually quite difficult to get rid of emotional memories, and at best, we can hope only to keep them under wraps. Studies by Maria A. Morgan in my laboratory have begun to illuminate how the brain regulates emotional expressions. Morgan has shown that when part of the prefrontal cortex is damaged, emotional memory is very hard to extinguish. This discovery indicates that the prefrontal areas—possibly by way of the amygdala—normally control expression of emotional memory and prevent emotional responses once they are no longer useful. A similar conclusion was proposed by Edmund T. Rolls and his colleagues at the University of Oxford during studies of primates. The researchers studied the electrical activity of neurons in the frontal cortex of the animals.

Functional variation in the pathway between this region of the cortex and the amygdala may make it more difficult for some people to change their emotional behavior. Davis and his colleagues have found that blocking NMDA receptors in the amygdala interferes with extinction. Those results hint that extinction is an active learning process. At the same time, such learning could be situated in connections between the prefrontal cortex and the amygdala. More experiments should disclose the answer.

Placing a basic emotional memory process in the amygdalic pathway yields obvious benefits. The amygdala is a critical site of learning because of its central location between input and output stations. Each route that leads to the amygdala—sensory thalamus, sensory cortex, and hippocampus—delivers unique information to the organ. Pathways originating in the sensory thalamus provide only a crude perception of the external world, but because they involve only one neural link, they are quite fast. In contrast, pathways from the cortex offer detailed and accurate representations, allowing us to recognize an object by sight or sound. But these pathways, which run from the thalamus to the sensory cortex to the amygdala, involve several neural links. And each link in the chain adds time.

Conserving time may be the reason there are two routes——one cortical and one subcortical—for emotional learning. Animals, and humans, need a quick-and-dirty reaction mechanism. The thalamus activates the amygdala at about the same time as it activates the cortex. The arrangement may enable emotional responses to begin in the amygdala before we completely recognize what it is we are reacting to or what we are feeling.

The thalamic pathway may be particularly useful in situations requiring a rapid response. Failing to respond to danger is more costly than responding inappropriately to a benign stimulus. For instance, the sound of rustling leaves is enough to alert us when we are walking in the woods without our having first to identify what

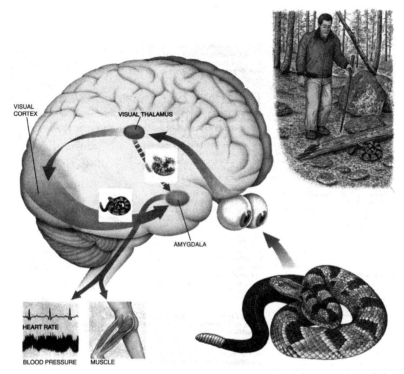

Cortical and subcortical pathways in the brain—generalized from our knowledge of the auditory system—may bring about a fearful response to a snake on a hiker's path. Visual stimuli are first processed by the thalamus, which passes rough, almost archetypal, information directly to the amygdala. This quick transmission allows the brain to start to respond to the possible danger. Meanwhile the visual cortex also receives information from the thalamus and, with more perceptual sophistication and more time, determines that there is a snake on the path. This information is relayed to the amygdala, causing heart rate and blood pressure to increase and muscles to contract. If, however, the cortex had determined that the object was not a snake, the message to the amygdala would quell the fear response. (Robert Osti)

is causing the sound. Similarly, the sight of a slender curved shape lying flat on the path ahead of us is sufficient to elicit defensive fear responses. We do not need to go through a detailed analysis of whether or not what we are seeing is a snake. Nor do we need to think about the fact that snakes are reptiles and that their skins can be used to make belts and boots. All these details are irrelevant and, in fact, detrimental to an efficient, speedy, and potentially lifesaving reaction. The brain simply needs to be able to store primitive cues and detect them. Later, coordination of this basic information with the cortex permits verification (yes, this is a snake) or brings the response (screaming, hyperventilation, or sprinting) to a stop.

Although the amygdala stores primitive information, we should not consider it the only learning center. The establishment of memories is a function of the entire network, not just of one component. The amygdala is certainly crucial, but we must not lose sight of the fact that its functions exist only by virtue of the system to which it belongs.

Memory is generally thought to be the process by which we bring back to mind some earlier conscious experience. The original learning and the remembering, in this case, are both conscious events. Workers have determined that declarative memory is mediated by the hippocampus and the cortex. But removal of the hippocampus has little effect on fear conditioning—except conditioning to context.

In contrast, emotional learning that comes about through fear conditioning is not declarative learning. Rather it is mediated by a different system, which in all likelihood operates independently of our conscious awareness. Emotional information may be stored within declarative memory, but it is kept there as a cold declarative fact. For example, if a person is injured in an automobile accident in which the horn gets stuck in the on position, he or she may later have a reaction when hearing the blare of car horns. The person may remember the details of the accident, such as where and when it occurred, who else was involved, and how awful it was. These are declarative memories that are dependent on the hippocampus. The individual may also become tense, anxious, and depressed, as the emotional memory is reactivated through the amygdalic system. The declarative system has stored the emotional content of the experience, but it has done so as a fact.

Emotional and declarative memories are stored and retrieved in parallel, and their activities are joined seamlessly in our conscious experience. That does not mean that we have direct conscious access to emotional memory; it means instead that we have access to the consequences—such as the way we behave, the way our bodies feel. These consequences combine with current declarative memory to form a new declarative memory. Emotion is not just unconscious memory: it exerts a powerful influence on declarative memory and other thought processes. As James L. McGaugh and his colleagues at the University of California at Irvine have convincingly shown, the amygdala plays an essential part in modulating the storage and strength of memories.

The distinction between declarative memory and emotional memory is an important one. W. J. Jacobs of the University of British Columbia and Lynn Nadel of the University of Arizona have argued that we are unable to remember traumatic events that take place early in life because the hippocampus has not yet matured to the point of forming consciously accessible memories. The emotional memory system, which may develop earlier, clearly forms and stores its unconscious memories of these events. And for this reason, the trauma may affect mental and behavioral functions in later life, albeit through processes that remain inaccessible to consciousness.

Because pairing a tone and a shock can bring about conditioned responses in animals throughout the phyla, it is clear that fear conditioning cannot be dependent on consciousness. Fruit flies and snails, for example, are not creatures known for their conscious mental processes. My way of interpreting this phenomenon is to consider fear a subjective state of awareness brought about when brain systems react to danger. Only if the organism possesses a sufficiently advanced neural mechanism does conscious fear accompany bodily response. This is not to say that only humans experience fear but, rather, that consciousness is a prerequisite to subjective emotional states.

Thus, emotions or feelings are conscious products of unconscious processes. It is crucial to remember that the subjective experiences we call feelings are not the primary business of the system that generates them. Emotional experiences are the result of triggering systems of behavioral adaptation that have been preserved by evolution. Subjective experience of any variety is challenging turf for scientists. We have, however, gone a long way toward understanding the neural system that underlies fear responses, and this same system may in fact give rise to subjective feelings of fear. If so, studies of the neural control of emotional responses may hold the key to understanding subjective emotion as well.

—June 1994

CREATING
FALSE MEMORIES

*Researchers are showing how suggestion and imagination
can create "memories" of events that did not actually occur*

Elizabeth F. Loftus

In 1986 Nadean Cool, a nurse's aide in Wisconsin, sought therapy from a psychiatrist to help her cope with her reaction to a traumatic event experienced by her daughter. During therapy, the psychiatrist used hypnosis and other suggestive techniques to dig out buried memories of abuse that Cool herself had allegedly experienced. In the process, Cool became convinced that she had repressed memories of having been in a satanic cult, of eating babies, of being raped, of having sex with animals, and of being forced to watch the murder of her eight-year-old friend. She came to believe that she had more than 120 personalities—children, adults, angels, and even a duck—all because, Cool was told, she had experienced severe childhood sexual and physical abuse. The psychiatrist also performed exorcisms on her, one of which lasted for five hours and included the sprinkling of holy water and screams for Satan to leave Cool's body.

When Cool finally realized that false memories had been planted, she sued the psychiatrist for malpractice. In March 1997, after five weeks of trial, her case was settled out of court for $2.4 million.

Nadean Cool is not the only patient to develop false memories as a result of questionable therapy. In Missouri in 1992 a church counselor helped Beth Rutherford to remember during therapy that her father, a clergyman, had regularly raped her between the ages of seven and 14 and that her mother sometimes helped him by holding her down. Under her therapist's guidance, Rutherford developed memories of her father twice impregnating her and forcing her to abort the fetus herself with a coat hanger. The father had to resign from his post as a clergyman when the allegations were made public. Later medical examination of the daughter revealed, however, that she was still a virgin at age 22 and had never been pregnant. The daughter sued the therapist and received a $1-million settlement in 1996.

About a year earlier two juries returned verdicts against a Minnesota psychiatrist accused of planting false memories by former patients Vynnette Hamanne and Elizabeth Carlson, who under hypnosis and sodium amytal, and after being fed misinformation about the workings of memory, had come to remember horrific abuse by family members. The juries awarded Hamanne $2.67 million and Carlson $2.5 million for their ordeals.

In all four cases, the women developed memories about childhood abuse in therapy and then later denied their authenticity. How can we determine if memories of childhood abuse are true or false? Without corroboration, it is very difficult to differentiate between false memories and true ones. Also, in these cases, some memories were contrary to physical evidence, such as explicit and detailed recollections of rape and abortion when medical examination confirmed virginity. How is it possible for people to acquire elaborate and confident false memories? A growing number of investigations demonstrate that under the right circumstances false memories can be instilled rather easily in some people.

My own research into memory distortion goes back to the early 1970s, when I began studies of the "misinformation effect." These studies show that when people who witness an event are later exposed to new and misleading information about it, their recollections often become distorted. In one example, participants viewed a simulated automobile accident at an intersection with a stop sign. After the viewing, half the participants received a suggestion that the traffic sign was a yield sign. When asked later what traffic sign they remembered seeing the intersection, those who had been given the suggestion tended to claim that they had seen a yield sign. Those who had not received the phony information were much more accurate in their recollection of the traffic sign.

My students and I have now conducted more than 200 experiments involving over 20,000 individuals that document how exposure to misinformation induces memory distortion. In these studies, people "recalled" a conspicuous barn in a bucolic scene that contained no buildings at all, broken glass and tape recorders that were not in the scenes they viewed, a white instead of a blue vehicle in a crime scene, and Minnie Mouse when they actually saw Mickey Mouse. Taken together, these studies show that misinformation can change an individual's recollections in predictable and sometimes very powerful ways.

Misinformation has the potential for invading our memories when we talk to other people, when we are suggestively interrogated or when we read or view media coverage about some event that we may have experienced ourselves. After more than two decades of exploring the power of misinformation, researchers have learned a great deal about the conditions that make people susceptible to memory modification. Memories are more easily modified, for instance, when the passage of time allows the original memory to fade.

■ FALSE CHILDHOOD MEMORIES

It is one thing to change a detail or two in an otherwise intact memory but quite another to plant a false memory of an event that never happened. To study false memory, my students and I first had to find a way to plant a pseudomemory that would not cause our subjects undue emotional stress, either in the process of creating the false memory or when we revealed that they had been intentionally deceived. Yet we wanted to try to plant a memory that would be at least mildly traumatic, had the experience actually happened.

My research associate, Jacqueline E. Pickrell, and I settled on trying to plant a specific memory of being lost in a shopping mall or large department store at about the age of five. Here's how we did it. We asked our subjects, 24 individuals ranging in age from 18 to 53, to try to remember childhood events that had been recounted to us by a parent, an older sibling, or another close relative. We prepared a booklet for each participant containing one-paragraph stories about three events that had actually happened to him or her and one that had not. We constructed the false event using information about a plausible shopping trip provided by a relative, who also verified that the participant had not in fact been lost at about the age of five. The lost-in-the-mall scenario included the following elements: lost for an extended period, crying, aid and comfort by an elderly woman, and, finally, reunion with the family.

After reading each story in the booklet, the participants wrote what they remembered about the event. If they did not remember it, they were instructed to write, "I do not remember this." In two follow-up interviews, we told the participants that we were interested in examining how much detail they could remember and how their memories compared with those of their relative. The event paragraphs were not read to them verbatim, but rather parts were provided as retrieval cues. The participants recalled something about 49 of the 72 true events (68 percent) immediately after the initial reading of the booklet and also in each of the two follow-up interviews. After reading the booklet, seven of the 24 participants (29 percent) remembered either partially or fully the false event constructed for them, and in the two follow-up interviews six participants (25 percent) continued to claim that they remembered the fictitious event. Statistically, there were some differences between the true memories and the false ones: participants used more words to describe the true memories, and they rated the true memories as being somewhat more clear. But if an onlooker were to observe many of our participants describe an event, it would be difficult indeed to tell whether the account was of a true or a false memory.

Of course, being lost, however frightening, is not the same as being abused. But the lost-in-the-mall study is not about real experiences of being lost; it is about planting false memories of being lost. The paradigm shows a way of instilling false memories and takes a step toward allowing us to understand how this might happen in real-world settings. Moreover, the study provides evidence that people can be led to remember their past in different ways, and they can even be coaxed into "remembering" entire events that never happened.

Studies in other laboratories using a similar experimental procedure have produced similar results. For instance, Ira Hyman, Troy H. Husband, and F. James Billing of Western Washington University asked college students to recall childhood experiences that had been recounted by their parents. The researchers told the students that the study was about how people remember shared experiences differently. In addition to actual events reported by parents, each participant was given one false event—either an overnight hospitalization for a high fever and a possible ear infection, or a birthday party with pizza and a clown—that supposedly happened at about the age of five. The parents confirmed that neither of these events actually took place.

Hyman found that students fully or partially recalled 84 percent of the true events in the first interview and 88 percent in the second interview. None of the participants recalled the false event during the first interview, but 20 percent said they remembered something about the false event in the second interview. One participant who had been exposed to the emergency hospitalization story later remembered a male doctor, a female nurse, and a friend from church who came to visit at the hospital.

In another study, along with true events Hyman presented different false events, such as accidentally spilling a bowl of punch on the parents of the bride at a wedding reception or having to evacuate a grocery store when the overhead sprinkler systems erroneously activated. Again, none of the participants recalled the false event during the first interview, but 18 percent remembered something about it in the second interview and 25 percent in the third interview. For example, during the first interview, one participant, when asked about the fictitious wedding event, stated, "I have no clue. I have never heard that one before." In the second interview, the participant said, "It was an outdoor wedding and I think we were running around and knocked something over like the punch bowl or something and made a big mess and of course got yelled at for it."

■ IMAGINATION INFLATION

The finding that an external suggestion can lead to the construction of false childhood memories helps us understand the process by which false memories arise. It is natural to wonder whether this research is applicable in real situations such as being interrogated by law officers or in psychotherapy. Although strong suggestion may not routinely occur in police questioning or therapy, suggestion in the form of an imagination exercise sometimes does. For instance, when trying to obtain a confession, law officers may ask a suspect to imagine having participated in a criminal act. Some mental health professionals encourage patients to imagine childhood events as a way of recovering supposedly hidden memories.

Surveys of clinical psychologists reveal that 11 percent instruct their clients to "let the imagination run wild," and 22 percent tell their clients to "give free rein to the imagination." Therapist Wendy Maltz, author of a popular book on childhood sexual abuse, advocates telling the patient: "Spend time imagining that you were sexually abused, without worrying about accuracy, proving anything, or having your ideas make sense. . . . Ask yourself . . . these questions: What time of day is it? Where are you? Indoors or outdoors? What kind of things are happening? Is there one or more person with you?" Maltz further recommends that therapists continue to ask questions such as "Who would have been likely perpetrators? When were you most vulnerable to sexual abuse in your life?"

The increasing use of such imagination exercises led me and several colleagues to wonder about their consequences. What happens when people imagine childhood experiences that did not happen to them? Does imagining a childhood event in-

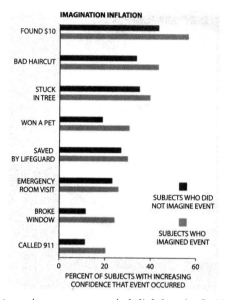

IMAGINATION INFLATION

FOUND $10

BAD HAIRCUT

STUCK
IN TREE

WON A PET

SAVED
BY LIFEGUARD

EMERGENCY
ROOM VISIT

BROKE
WINDOW

CALLED 911

SUBJECTS WHO DID
NOT IMAGINE EVENT

SUBJECTS WHO
IMAGINED EVENT

0   20   40   60
PERCENT OF SUBJECTS WITH INCREASING
CONFIDENCE THAT EVENT OCCURRED

Imagining an event can increase a person's belief that the fictitious event actually happened. To study the "imagination inflation" effect, the author and her colleagues asked participants to indicate on a scale the likelihood that each of forty events occurred during their childhood. Two weeks later they were given guidance in imagining some of the events they said had not taken place and then were asked to rate the original forty events again. Whereas all participants showed increased confidence that the events had occurred, those who took part in actively imagining the events reported an even greater increase. (Bryan Christie)

crease confidence that it occurred? To explore this, we designed a three-stage procedure. We first asked individuals to indicate the likelihood that certain events happened to them during their childhood. The list contains 40 events, each rated on a scale ranging from "definitely did not happen" to "definitely did happen." Two weeks later we asked the participants to imagine that they had experienced some of these events. Different subjects were asked to imagine different events. Sometime later the participants again were asked to respond to the original list of 40 childhood events, indicating how likely it was that these events actually happened to them.

Consider one of the imagination exercises. Participants are told to imagine playing inside at home after school, hearing a strange noise outside, running toward the window, tripping, falling, reaching out, and breaking the window with their hand. In addition, we asked participants questions such as "What did you trip on? How did you feel?"

In one study 24 percent of the participants who imagined the broken-window scenario later reported an increase in confidence that the event had occurred, whereas only 12 percent of those who were not asked to imagine the incident reported an increase in the likelihood that it had taken place. We found this "imagination inflation" effect in each of the eight events that participants were asked to imagine. A number of possible explanations come to mind. An obvious one is that an act of imagination simply makes the event seem more familiar and that familiarity is mistakenly related to childhood memories rather than to the act of imagination. Such source confusion—when a person does not remember the source of information—can be especially acute for the distant experiences of childhood.

Studies by Lyn Goff and Henry L. Roediger III of Washington University of recent rather than childhood experiences more directly connect imagined actions to the construction of false memory. During the initial session, the researchers instructed participants to perform the stated action, imagine doing it, or just listen to the statement and do nothing else. The actions were simple ones: knock on the table, lift the stapler, break the toothpick, cross your fingers, roll your eyes. During the second session, the participants were asked to imagine some of the actions that they had not previously performed. During the final session, they answered questions about what actions they actually performed during the initial session. The investigators found that the more times participants imagined an unperformed action, the more likely they were to remember having performed it.

■ IMPOSSIBLE MEMORIES

It is highly unlikely that an adult can recall genuine episodic memories from the first year of life, in part because the hippocampus, which plays a key role in the creation of memories, has not matured enough to form and store long-lasting memories that can be retrieved in adulthood. A procedure for planting "impossible" memories about experiences that occur shortly after birth has been developed by the late Nicholas Spanos and his collaborator at Carleton University. Individuals are led to believe that they have well-coordinated eye movements and visual exploration skills probably because they were born in hospitals that hung swinging, colored mobiles over infant cribs. To confirm whether they had such an experience, half the participants are hypnotized, age-regressed to the day after birth and asked what they remembered. The other half of the group participates in a "guided mnemonic restructuring" procedure that uses age regression as well as active encouragement to re-create the infant experiences by imagining them.

Spanos and his co-workers found that the vast majority of their subjects were susceptible to these memory-planting procedures. Both the hypnotic and guided participants reported infant memories. Surprisingly, the guided group did so somewhat more (95 versus 70 percent). Both groups remembered the colored mobile at a relatively high rate (56 percent of the guided group and 46 percent of the hypnotic subjects). Many participants who did not remember the mobile did recall other things, such as doctors, nurses, bright lights, cribs, and masks. Also, in both groups, of those who reported memories of infancy, 49 percent felt that they were real memories, as opposed to 16 percent who claimed that they were merely fantasies. These findings confirm earlier studies that many individuals can be led to construct complex, vivid, and detailed false memories via a rather simple procedure. Hypnosis clearly is not necessary.

■ HOW FALSE MEMORIES FORM

In the lost-in-the-mall study, implantation of false memory occurred when another person, usually a family member, claimed that the incident happened. Corroboration of an event by another person can be a powerful technique for instilling a false memory. In fact, merely claiming to have seen a person do something can lead that person to make a false confession of wrongdoing.

This effect was demonstrated in a study by Saul M. Kassin and his colleagues at Williams College, who investigated the reactions of individuals falsely accused of damaging a computer by pressing the wrong key. The innocent participants initially denied the charge, but when a confederate said that she had seen them perform the action, many participants signed a confession, internalized guilt of the act, and went on to confabulate details that were consistent with that belief. These findings show that false incriminating evidence can induce people to accept guilt for a crime they did not commit and even to develop memories to support their guilty feelings.

Research is beginning to give us an understanding of how false memories of complete, emotional, and self-participatory experiences are created in adults. First, there are social demands on individuals to remember; for instance, researchers exert some pressure on participants in a study to come up with memories. Second, memory construction by imagining events can be explicitly encouraged when people are having trouble remembering. And, finally, individuals can be encouraged not to think about whether their constructions are real or not. Creation of false memories is most likely to occur when these external factors are present, whether in an experimental setting, in a therapeutic setting, or during everyday activities.

False memories are constructed by combining actual memories with the content of suggestions received from others. During the process, individuals may forget the source of the information. This is a classic example of source confusion, in which the content and the source become dissociated.

Of course, because we can implant false childhood memories in some individuals in no way implies that all memories that arise after suggestion are necessarily false. Put another way, although experimental work on the creation of false memories may raise doubt about the validity of long-buried memories, such as repeated trauma, it in no way disproves them. Without corroboration, there is little that can be done to help even the most experienced evaluator to differentiate true memories from ones that were suggestively planted.

The precise mechanisms by which such false memories are constructed await further research. We still have much to learn about the degree of confidence and the characteristics of false memories created in these ways, and we need to discover what types of individuals are particularly susceptible to these forms of suggestion and who is resistant.

As we continue this work, it is important to heed the cautionary tale in the data we have already obtained: mental health professionals and others must be aware of how greatly they can influence the recollection of events and of the urgent need for maintaining restraint in situations in which imagination is used as an aid in recovering presumably lost memories.

—September 1997

THE SPLIT BRAIN
REVISITED

Groundbreaking work that began more than
a quarter of a century ago has led to ongoing insights
about brain organization and consciousness

Michael S. Gazzaniga

About 30 years ago in these very pages, I wrote about dramatic new studies of the brain. Three patients who were seeking relief from epilepsy had undergone surgery that severed the corpus callosum—the superhighway of neurons connecting the halves of the brain. By working with these patients, my colleagues Roger W. Sperry, Joseph E. Bogen, P. J. Vogel, and I witnessed what happened when the left and the right hemispheres were unable to communicate with each other.

It became clear that visual information no longer moved between the two sides. If we projected an image to the right visual field—that is, to the left hemisphere, which is where information from the right field is processed—the patients could describe what they saw. But when the same image was displayed to the left visual field, the patients drew a blank: they said they didn't see anything. Yet if we asked them to point to an object similar to the one being projected, they could do so with ease. The right brain saw the image and could mobilize a nonverbal response. It simply couldn't talk about what it saw.

The same kind of finding proved true for touch, smell, and sound. Additionally, each half of the brain could control the upper muscles of both arms, but the muscles manipulating hand and finger movement could be orchestrated only by the contralateral hemisphere. In other words, the right hemisphere could control only the left hand and the left hemisphere only the right hand.

Ultimately, we discovered that the two hemispheres control vastly different aspects of thought and action. Each half has its own specialization and thus its own limitations and advantages. The left brain is dominant for language and speech. The right excels at visual-motor tasks. The language of these findings has become part of our culture: writers refer to themselves as left-brained, visual artists as right-brained.

In the intervening decades, split-brain research has continued to illuminate many areas of neuroscience. Not only have we and others learned even more about how the hemispheres differ, but we also have been able to understand how they communicate once they have been separated. Split-brain studies have shed light on language, on mechanisms of perception and attention, and on brain organization as well as the potential seat of false memories. Perhaps most intriguing has been the contribution of these studies to our understanding of consciousness and evolution.

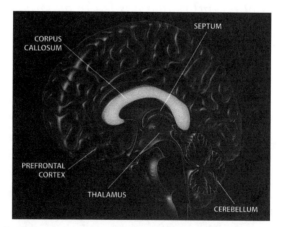

Brain wiring is, in many cases, contralateral. The right hemisphere processes information from the left visual field, whereas the left hemisphere processes data from the right visual field. For hand movement as well, the right hemisphere controls the hand and fingers of the left arm; the left hemisphere controls the right. Both hemispheres, however, dictate the movement of the upper arms. The two hemispheres are connected by neuronal bridges called commissures. The largest of these, and the one severed during split-brain operations, is the corpus callosum. (John W. Karapelou)

The original split-brain studies raised many interesting questions, including ones about whether the distinct halves could still "talk" to each other and what role any such communication played in thought and action. There are several bridges of neurons, called commissures, that connect the hemispheres. The corpus callosum is the most massive of these and typically the only one severed during surgery for epilepsy. But what of the many other, smaller commissures?

■ REMAINING BRIDGES

By studying the attentional system, researchers have been able to address this question. Attention involves many structures in the cortex and the subcortex—the oldest, more primitive part of our brains. In the 1980s Jeffrey D. Holtzman of Cornell University Medical College found that each hemisphere is able to direct spatial attention not only to its own sensory sphere but also to certain points in the sensory sphere of the opposite, disconnected hemisphere. This discovery suggests that the attentional system is common to both hemispheres—at least with regard to spatial information—and can still operate via some remaining interhemispheric connections.

Holtzman's work was especially intriguing because it raised the possibility that there were finite attentional "resources." He posited that working on one kind of task uses certain brain resources; the harder the task, the more of these resources are needed—and the more one half of the brain must call on the subcortex or the other hemisphere for help. In 1982 Holtzman led the way again, discovering that, indeed, the harder one half of a split brain worked, the harder it was for the other half to carry out another task simultaneously.

Recent investigations by Steve J. Luck of the University of Iowa, Steven A. Hillyard and his colleagues at the University of California at San Diego, and Ronald Mangun of the University of California at Davis show that another aspect of attention is also preserved in the split brain. They looked at what happens when a person searches a visual field for a pattern or an object. The researchers found that split-brain patients perform better than normal people do in some of these visual-searching tasks. The intact brain appears to inhibit the search mechanisms that each hemisphere naturally possesses.

The left hemisphere, in particular, can exert powerful control over such tasks. Alan Kingstone of the University of Alberta found that the left hemisphere is "smart" about its search strategies, whereas the right is not. In tests where a person can deduce how to search efficiently an array of similar items for an odd exception, the left does better than the right. Thus, it seems that the more competent left hemisphere can hijack the intact attentional system.

TESTING FOR SYNTHESIS

Abililty to synthesize information is lost after split-brain surgery, as this experiment shows. One hemisphere of a patient was flashed a card with the word "bow"; the other hemisphere saw "arrow." Because the patient drew a bow and arrow, my colleagues and I assumed the two hemispheres were still able to communicate with each other—despite the severing of the corpus callosum—and had integrated the words into a meaningful composite.

The next test proved us wrong. We flashed "sky" to one hemisphere, "scraper" to the other. The resulting image revealed that the patient was not synthesizing information: sky atop a comblike scraper was drawn, rather than a tall building. One hemisphere drew what it had seen, then the other drew its word. In the case of bow and arrow, the superposition of the two images misled us because the picture appeared integrated. Finally, we tested to see whether each hemisphere could, on its own, integrate words. We flashed "fire" and then "arm" to the right hemisphere. The left hand drew a rifle rather than an arm on fire, so it was clear that each hemisphere was capable of synthesis.

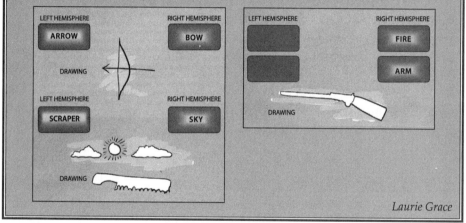

Laurie Grace

Although these and other studies indicated that some communication between the split hemispheres remains, other apparent interhemispheric links proved illusory. I conducted an experiment with Kingstone, for instance, that nearly misled us on this front. We flashed two words to a patient and then asked him to draw what he saw. "Bow" was flashed to one hemisphere and "arrow" to the other. To our surprise, our patient drew a bow and arrow! It appeared as though he had internally integrated the information in one hemisphere; that hemisphere had, in turn, directed the drawn response.

We were wrong. We finally determined that integration had actually taken place on the paper, not in the brain. One hemisphere had drawn its item—the bow—and then the other hemisphere had gained control of the writing hand, drawing its stimulus—the arrow—on top of the bow. The image merely looked coordinated. We discovered this chimera by giving less easily integrated word pairs like "sky" and "scraper." The subject did not draw a tall building; instead he drew the sky over a picture of a scraper.

■ THE LIMITS OF EXTRAPOLATION

In addition to helping neuroscientists determine which systems still work and which are severed along with the corpus callosum, studies of communication between the hemispheres led to an important finding about the limits of nonhuman studies. Humans often turn to the study of animals to understand themselves. For many years, neuroscientists have examined the brains of monkeys and other creatures to explore the ways in which the human brain operates. Indeed, it has been a common belief—emphatically disseminated by Charles Darwin—that the brains of our closest relatives have an organization and function largely similar, if not identical, to our own.

Split-brain research has shown that this assumption can be spurious. Although some structures and functions are remarkably alike, differences abound. The anterior commissure provides one dramatic example. This small structure lies somewhat below the corpus callosum. When this commissure is left intact in otherwise split-brain monkeys, the animals retain the ability to transfer visual information from one hemisphere to the other. People, however, do not transfer visual information in any way. Hence, the same structure carries out different functions in different species—an illustration of the limits of extrapolating from one species to another.

Even extrapolating between people can be dangerous. One of our first striking findings was that the left brain could freely process language and speak about its experience. Although the right was not so free, we also found that it could process some language. Among other skills, the right hemisphere could match words to pictures, do spelling and rhyming, and categorize objects. Although we never found any sophisticated capacity for syntax in that half of the brain, we believed the extent of its lexical knowledge to be quite impressive.

Over the years it has become clear that our first three cases were unusual. Most people's right hemispheres cannot handle even the most rudimentary language, contrary to what we initially observed. This finding is in keeping with other neurological data, particularly those from stroke victims. Damage to the left hemisphere is far more detrimental to language function than is damage to the right.

Nevertheless, there exists a great deal of plasticity and individual variation. One patient, dubbed J. W., developed the capacity to speak out of the right hemisphere—13 years after surgery. J. W. can now speak about information presented to the left or to the right brain.

Kathleen B. Baynes of the University of California at Davis reports another unique case. A left-handed patient spoke out of her left brain after split-brain surgery—not a surprising finding in itself. But the patient could *write* only out of her right, nonspeaking hemisphere. This dissociation confirms the idea that the capacity to write need not be associated with the capacity for phonological representation. Put differently, writing appears to be an independent system, an intention of the human species. It can stand alone and does not need to be part of our inherited spoken language system.

■ BRAIN MODULES

Despite myriad exceptions, the bulk of split-brain research has revealed an enormous degree of lateralization—that is, specialization in each of the hemispheres. As investigators have struggled to understand how the brain achieves its goals and how it is organized, the lateralization revealed by split-brain studies has figured into what is called the modular model. Research in cognitive science, artificial intelligence, evolutionary psychology, and neuroscience has directed attention to the idea that brain and mind are built from discrete units—or modules—that carry out specific functions. According to this theory, the brain is not a general problem-solving device whose every part is capable of any function. Rather it is a collection of devices that assists the mind's information-processing demands.

Within that modular system, the left hemisphere has proved quite dominant for major cognitive activities, such as problem solving. Split-brain surgery does not seem to affect these functions. It is as if the left hemisphere has no need for the vast computational power of the other half of the brain to carry out high-level activities. The right hemisphere, meanwhile, is severely deficient in difficult problem solving.

Joseph E. LeDoux of New York University and I discovered this quality of the left brain almost 20 years ago. We had asked a simple question: How does the left hemisphere respond to behaviors produced by the silent right brain? Each hemisphere was presented a picture that related to one of four pictures placed in front of the split-brain subject. The left and the right hemispheres easily picked the correct card. The left hand pointed to the right hemisphere's choice and the right hand to the left hemisphere's choice [*see illustration on page 135*].

We then asked the left hemisphere—the only one that can talk—why the left hand was pointing to the object. It really did not know, because the decision to point to the card was made in the right hemisphere. Yet, quick as a flash, it made up

FINDING FALSE MEMORY

False memories originate in the left hemisphere. As MRI images have indicated, a region in both the right and left hemispheres is active when a false memory is recalled, but only the right is active during a true memory. My colleagues and I studied this phenomenon by testing the narrative ability of the left hemisphere. Each hemisphere was shown four small pictures, one of which related to a larger picture also presented to that hemisphere. The patient had to choose the most appropriate small picture.

As seen below, the right hemisphere—that is, the left hand—correctly picked the shovel for the snowstorm; the right hand, controlled by the left hemisphere, correctly picked the chicken to go with the bird's foot. Then we asked the patient why the left hand—or right hemisphere—was pointing to the shovel. Because only the left hemisphere retains the ability to talk, it answered. But because it could not know why the right hemisphere was doing what it was doing, it made up a story about what it could see—namely, the chicken. It said the right hemisphere chose the shovel to clean out a chicken shed.

John W. Karapelou

an explanation. We dubbed this creative, narrative talent the interpreter mechanism.

This fascinating ability has been studied recently to determine how the left hemisphere interpreter affects memory. Elizabeth A. Phelps of Yale University, Janet Metcalfe of Columbia University, and Margaret Funnell, a postdoctoral fellow at Dartmouth College, have found that the two hemispheres differ in their ability to process new data. When presented with new information, people usually remember much of what they experience. When questioned, they also usually claim to remember things that were not truly part of the experience. If split-brain patients are given such tests, the left hemisphere generates many false reports. But the right brain does not; it provides a much more veridical account.

This finding may help researchers determine where and how false memories develop. There are several views about when in the cycle of information processing such memories are laid down. Some researchers suggest they develop early in the cycle, that erroneous accounts are actually encoded at the time of the event. Others believe false memories reflect an error in reconstructing past experience: in other words, that people develop a schema about what happened and retrospectively fit untrue events—that are nonetheless consistent with the schema—into their recollection of the original experience.

The left hemisphere has exhibited certain characteristics that support the latter view. First, developing such schemata is exactly what the left hemisphere interpreter excels at. Second, Funnell has discovered that the left hemisphere has an ability to determine the source of a memory, based on the context or the surrounding events. Her work indicates that the left hemisphere actively places its experiences in a larger context, whereas the right simply attends to the perceptual aspects of the stimulus. Finally, Michael B. Miller, a graduate student at Dartmouth, has demonstrated that the left prefrontal regions of normal subjects are activated when they recall false memories.

These findings all suggest that the interpretive mechanism of the left hemisphere is always hard at work, seeking the meaning of events. It is constantly looking for order and reason, even when there is none—which leads it continually to make mistakes. It tends to overgeneralize, frequently constructing a potential past as opposed to a true one.

■ THE EVOLUTIONARY PERSPECTIVE

George L. Wolford of Dartmouth has lent even more support to this view of the left hemisphere. In a simple test that requires a person to guess whether a light is going to appear on the top or bottom of a computer screen, humans perform inventively. The experimenter manipulates the stimulus so that the light appears on the top 80 percent of the time but in a random sequence. While it quickly becomes evident that the top button is being illuminated more often, people invariably try to figure out the entire pattern or sequence—and they deeply believe they can. Yet by adopting this strategy, they are correct only 68 percent of the time. If they always pressed the top button, they would be correct 80 percent of the time.

Rats and other animals, on the other hand, are more likely to "learn to maximize" and to press only the top button. It turns out the right hemisphere behaves in the same way: it does not try to interpret its experience and find deeper meaning. It continues to live only in the thin moment of the present—and to be correct 80 percent of the time. But the left, when asked to explain why it is attempting to figure the whole sequence, always comes up with a theory, no matter how outlandish.

This narrative phenomenon is best explained by evolutionary theory. The human brain, like any brain, is a collection of neurological adaptations established through natural selection. These adaptations each have their own representation—that is, they can be lateralized to specific regions or networks in the brain. Throughout the animal kingdom, however, capacities are generally not lateralized. Instead they tend to be found in both hemispheres to roughly equal degrees. And although monkeys show some signs of lateral specialization, these are rare and inconsistent.

For this reason, it has always appeared that the lateralization seen in the human brain was an evolutionary add-on—mechanisms or abilities that were laid down in one hemisphere only. We recently stumbled across an amazing hemispheric dissociation that challenges this view. It forced us to speculate that some lateralized phenomena may arise from a hemisphere's losing an ability—not gaining it.

In what must have been fierce competition for cortical space, the evolving primate brain would have been hard-pressed to gain new faculties without losing old ones. Lateralization could have been its salvation. Because the two hemispheres are connected, mutational tinkering with a homologous cortical region could give rise to a new function—yet not cost the animal, because the other side would remain unaffected.

Paul M. Corballis, a postdoctoral fellow at Dartmouth, and Robert Fendrich of Dartmouth, Robert M. Shapley of New York University, and I studied in many split-brain patients the perception of what are called illusory contours. Earlier work had suggested that seeing the well-known illusory contours of Gaetano Kanizsa of the University of Trieste was the right hemisphere's specialty. Our experiments revealed a different situation.

We discovered that both hemispheres could perceive illusory contours—but that the right hemisphere was able to grasp certain perceptual groupings that the left could not. Thus, while both hemispheres in a split-brain person can judge whether the illusory rectangles are fat or thin when no line is drawn around the openings of the "Pacman" figures, only the right can continue to make the judgment after the line has been drawn [see illustration on page 138]. This setup is called the amodal version of the test.

What is so interesting is that Kanizsa himself has demonstrated that mice can do the amodal version. That a lowly mouse can perceive perceptual groupings, whereas a human's left hemisphere cannot, suggests that a capacity has been lost. Could it be that the emergence of a human capacity like language—or an interpretive mechanism—chased this perceptual skill out of the left brain? We think so, and

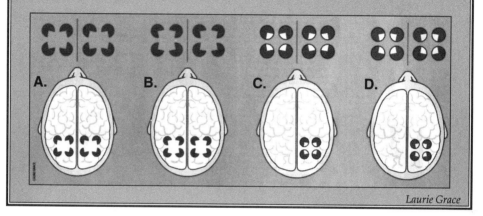

LOOKING FOR ILLUSIONS

Illusory contours reveal that the human right brain can process some things the left cannot. Both hemispheres can "see" whether the illusory rectangles of this experiment are fat (*a*) or thin (*b*). But when outlines are added, only the right brain can still tell the difference (*c* and *d*). In mice, however, both hemispheres can consistently perceive these differences. For a rodent to perform better than we do suggests that some capabilities were lost from one hemisphere or the other as the human brain evolved. New capabilities may have squeezed out old ones in a race for space.

Laurie Grace

this opinion gives rise to a fresh way of thinking about the origins of lateral specialization.

Our uniquely human skills may well be produced by minute and circumscribed neuronal networks. And yet our highly modularized brain generates the feeling in all of us that we are integrated and unified. How so, given that we are a collection of specialized modules?

The answer may be that the left hemisphere seeks explanations for why events occur. The advantage of such a system is obvious. By going beyond the simple observation of events and asking why they happened, a brain can cope with these same events better, should they happen again.

Realizing the strengths and weaknesses of each hemisphere prompted us to think about the basis of mind, about this overarching organization. After many years of fascinating research on the split brain, it appears that the inventive and interpreting left hemisphere has a conscious experience very different from that of the truthful, lateral right brain. Although both hemispheres can be viewed as conscious, the left brain's consciousness far surpasses that of the right. Which raises another set of questions that should keep us busy for the next 30 years or so.

—July 1998

The Biological Basis of Learning and Individuality

*Recent discoveries suggest that learning engages
a simple set of rules that modify the strength of connections
between neurons in the brain. These changes play
an important role in making each individual unique*

Eric R. Kandel and Robert D. Hawkins

Over the past several decades, there has been a gradual merger of two originally separate fields of science: neurobiology, the science of the brain, and cognitive psychology, the science of the mind. Recently the pace of unification has quickened, with the result that a new intellectual framework has emerged for examining perception, language, memory, and conscious awareness. This new framework is based on the ability to study the biological substrates of these mental functions. A particularly fascinating example can be seen in the study of learning. Elementary aspects of the neuronal mechanisms important for several different types of learning can now be studied on the cellular and even on the molecular level. The analysis of learning may therefore provide the first insights into the molecular mechanisms underlying a mental process and so begin to build a bridge between cognitive psychology and molecular biology.

Learning is the process by which we acquire new knowledge, and memory is the process by which we retain that knowledge over time. Most of what we know about the world and its civilizations we have learned. Thus, learning and memory are central to our sense of individuality. Indeed, learning goes beyond the individual to the transmission of culture from generation to generation. Learning is a major vehicle for behavioral adaptation and a powerful force for social progress. Conversely, loss of memory leads to loss of contact with one's immediate self, with one's life history, and with other human beings.

Until the middle of the 20th century, most students of behavior did not believe that memory was a distinct mental function independent of movement, perception, attention, and language. Long after those functions had been localized to different regions of the brain, researchers still doubted that memory could ever be assigned to a specific region. The first person to do so was Wilder G. Penfield, a neurosurgeon at the Montreal Neurological Institute.

In the 1940s Penfield began to use electrical stimulation to map motor, sensory, and language functions in the cortex of patients undergoing neurosurgery for the relief of epilepsy. Because the brain itself does not have pain receptors, brain surgery can be carried out under local anesthesia in fully conscious patients, who can describe what they experience in response to electric stimuli applied to different cortical areas. Penfield explored the cortical surface in more than 1,000 patients. Occasionally he found that electrical stimulation produced an experiential response, or flashback, in which the patients described a coherent recollection of an earlier experience. These memorylike responses were invariably elicited from the temporal lobes.

Additional evidence for the role of the temporal lobe in memory came in the 1950s from the study of a few patients who underwent bilateral removal of the hippocampus and neighboring regions in the temporal lobe as treatment for epilepsy. In the first and best-studied case, Brenda Milner of the Montreal Neurological Institute described a 27-year-old assembly-line worker, H.M., who had suffered from untreatable and debilitating temporal lobe seizures for more than 10 years. The surgeon William B. Scoville removed the medial portion of the temporal lobes on both sides of H.M.'s brain. The seizure disturbance was much improved. But immediately after the operation, H.M. experienced a devastating memory deficit: he had lost the capacity to form new long-term memories.

Despite his difficulty with the formation of new memories, H.M. still retained his previously acquired long-term memory store. He remembered his name, retained a perfectly good use of language, and kept his normal vocabulary; his IQ remained in the range of bright-normal. He remembered well the events that preceded the surgery, such as the job he had held, and he remembered vividly the events of his childhood. Moreover, H.M. still had a completely intact short-term memory. What H.M. lacked, and lacked profoundly, was the ability to translate what he learned from short-term to long-term memory. For example, he could converse normally with the hospital staff, but he did not remember them even though he saw them every day.

The memory deficit following bilateral temporal lobe lesions was originally thought to apply equally to all forms of new learning. But Milner soon discovered that this is not the case. Even though patients with such lesions have profound deficits, they can accomplish certain types of learning tasks as well as normal subjects can and retain the memory of these tasks for long periods. Milner first demonstrated this residual memory capability in H.M. with the discovery that he could learn new motor skills normally. She, and subsequently Elizabeth K. Warrington of the National Hospital for Nervous Diseases in London and Lawrence Weiskrantz of the University of Oxford, found that patients such as H.M. can also acquire and retain memory for elementary kinds of learning that involve changing the strength of reflex responses, such as habituation, sensitization, and classical conditioning.

It immediately became apparent to students of behavior that the difference between types of learning that emerged from studies of patients with temporal lobe lesions represented a fundamental psychological distinction—a division in the way all of us acquire knowledge. Although it is still not clear how many distinct memory systems there are, researchers agree that lesions of the temporal lobes severely impair forms of learning and memory that require a conscious record. In accordance with the suggestion of Neal J. Cohen of the University of Illinois and Larry R. Squire of the University of California at San Diego and of Daniel L. Schacter of the University of Toronto, these types of learning are commonly called declarative or explicit. Those forms of learning that do not utilize conscious participation remain surprisingly intact in patients with temporal lobe lesions; they are referred to as nondeclarative or implicit.

Explicit learning is fast and may take place after only one training trial. It often involves association of simultaneous stimuli and permits storage of information about a single event that happens in a particular time and place; it therefore affords a sense of familiarity about previous events. In contrast, implicit learning is slow and accumulates through repetition over many trials. It often involves association of sequential stimuli and permits storage of information about predictive relations between events. Implicit learning is expressed primarily by improved performance of certain tasks without the subject being able to describe just what has been learned, and it involves memory systems that do not draw on the contents of the general knowledge of the individual. When a subject such as H.M. is asked why he performs a given task better after five days of practice than on the first day, he may respond, "What are you talking about? I've never done this task before."

Whereas explicit memory requires structures in the temporal lobe of vertebrates, implicit memory is thought to be expressed through activation of the particular sensory and motor systems engaged by the learning task; it is acquired and retained by the plasticity inherent in these neuronal systems. As a result, implicit memory can be studied in various reflex systems in either vertebrates or invertebrates. Indeed, even simple invertebrate animals show excellent reflexive learning.

The existence of two distinct forms of learning has caused the reductionists among neurobiologists to ask whether there is a representation on the cellular level for each of these two types of learning process. Both the neural systems that mediate explicit memory and those that mediate implicit memory can store information about the association of stimuli. But does the same set of cellular learning rules guide the two memory systems as they store associations, or do separate sets of rules govern each system?

An assumption underlying early studies of the neural basis of memory systems was that the storage of associative memory, both implicit and explicit, required a fairly complex neural circuit. One of the first to challenge this view was the Canadian psychologist Donald O. Hebb, a teacher of Milner. Hebb boldly suggested that associative learning could be produced by a simple cellular mechanism. He proposed that associations could be formed by coincident neural activity: "When an axon of cell A . . . excite[s] cell B and repeatedly or persistently takes part in firing it, some growth process or metabolic change takes place in one or both cells such that A's efficacy, as one of the cells firing B, is increased." According to Hebb's learning rule, coincident activity in the presynaptic and postsynaptic neurons is critical for strengthening the connection between them (a so-called pre-post associative mechanism) [*see illustration on page 143*].

Ladislav Tauc and one of us (Kandel) proposed a second associative learning rule in 1963 while working at the Institute Marey in Paris on the nervous system of the marine snail *Aplysia*. They found that the synaptic connection between two neurons could be strengthened without activity of the postsynaptic cell when a third neuron acts on the presynaptic neuron. The third neuron, called a modulatory neuron, enhances transmitter release from the terminals of the presynaptic neuron. They suggested that this mechanism could take on associative properties if the electrical impulses known as action potentials in the presynaptic cell were coincident with action potentials in the modulatory neuron (a premodulatory associative mechanism).

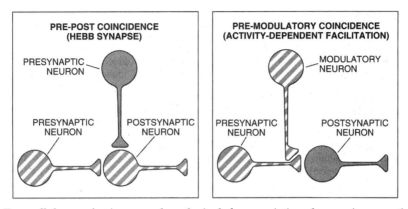

Two cellular mechanisms are hypothesized for associative changes in synaptic strength during learning. The pre-post coincidence mechanism, proposed by Donald O. Hebb in 1949, posits that coincident activity in the presynaptic and postsynaptic neurons is critical for strengthening the connections between them. The premodulatory coincidence mechanism proposed in 1963, based on studies in Aplysia, holds that the connection can be strengthened without activity of the postsynaptic cell when a third neuron, the modulatory neuron, is active at the same time as the presynaptic neuron. Stripes denote neurons in which coincident activity must occur to produce the associative change. (Ian Worpole)

Subsequently, we and our colleagues Thomas J. Carew and Thomas W. Abrams of Columbia University and Edgar T. Walters and John H. Byrne of the University of Texas Health Science Center found experimental confirmation. We observed the pre-modulatory associative mechanism in *Aplysia,* where it contributes to classical conditioning, an implicit form of learning. Then, in 1986, Holger J. A. Wigström and Bengt E. W. Gustafsson, working at the University of Göteborg, found that the pre-post associative mechanism occurs in the hippocampus, where it is utilized in types of synaptic change that are important for spatial learning, an explicit form of learning.

The finding of two distinct cellular learning rules, each with associative properties, suggested that the associative mechanisms for implicit and explicit learning need not require complex neural networks. Rather the ability to detect associations may simply reflect the intrinsic capability of certain cellular interactions. Moreover, these findings raised an intriguing question: Are these apparently different mechanisms in any way related? Before considering their possible interrelation, we shall first describe the two learning mechanisms, beginning with the pre-modulatory mechanism contributing to classical conditioning in *Aplysia.*

Classical conditioning was first described at the turn of the century by the Russian physiologist Ivan Pavlov, who immediately appreciated that conditioning represents the simplest example of learning to associate two events. In classical conditioning, an ineffective stimulus called the conditioned stimulus (or more correctly the to-be-conditioned stimulus) is repeatedly paired with a highly effective stimulus called the unconditioned stimulus. The conditioned stimulus initially produces only a small response or no response at all; the unconditioned stimulus elicits a powerful response without requiring prior conditioning.

As a result of conditioning (or learning), the conditioned stimulus becomes capable of producing either a larger response or a completely new response. For example, the sound of a bell (the conditioned stimulus) becomes effective in eliciting a behavioral response such as lifting a leg only after that sound has been paired with a shock to the leg (the unconditioned stimulus) that invariably produces a leg-lifting response. For conditioning to occur, the conditioned stimulus generally must be correlated with the unconditioned stimulus and precede it by a certain critical period. The animal is therefore thought to learn predictive relations between the two stimuli.

Because *Aplysia* has a nervous system containing only about 20,000 central nerve cells, aspects of classical conditioning can be examined at the cellular level. *Aplysia* has a number of simple reflexes of which the gill-withdrawal reflex has been particularly well studied. The animal normally withdraws the gill, its respiratory organ, when a stimulus is applied to another part of its body such as the mantle shelf or the fleshy extension called the siphon. Both the mantle shelf and the siphon are innervated by their own populations of sensory neurons. Each of these populations makes direct contact with motor neurons for the gill as well as with various classes of excitatory and inhibitory interneurons that synapse on the motor neurons. We and our colleagues Carew and Walters found that even this simple reflex can be conditioned.

A weak tactile stimulus to one pathway, for example, the siphon, can be paired with an unconditioned stimulus (a strong shock) to the tail. The other pathway, the mantle shelf, can then be used as a control pathway. The control pathway is stimulated the same number of times, but the stimulus is not paired (associated) with the tail shock. After five pairing trials, the response to stimulation of the siphon (the paired pathway) is greater than that of the mantle (the unpaired pathway). If the procedure is reversed and the mantle shelf is paired rather than the siphon, the response to the mantle shelf will be greater than that to the siphon. This differential conditioning is remarkably similar in several respects to that seen in vertebrates.

To discover how this conditioning works, we focused on one component: the connections between the sensory neurons and their target cells, the interneurons and motor neurons. Stimulating the sensory neurons from either the siphon or the mantle shelf generates excitatory synaptic potentials in the interneurons and motor cells. These synaptic potentials cause the motor cells to discharge, leading to a brisk reflex withdrawal of the gill. The unconditioned reinforcing stimulus to the tail activates many cell groups, some of which also cause movement of the gill. Among them are at least three groups of modulatory neurons, in one of which the chemical serotonin is the transmitter. (Neurotransmitters such as serotonin that carry messages between cells are called first messengers; other chemicals known as second messengers relay information within the cell.)

These modulatory neurons act on the sensory neurons from both the siphon and the mantle shelf, where they produce presynaptic facilitation; that is, they enhance transmitter release from the terminals of the sensory neurons. Presynaptic facilitation contributes to a nonassociative form of learning called sensitization, in which an animal learns to enhance a variety of defensive reflex responses after receiving a noxious stimulus. This type of learning is referred to as nonassociative because it does not depend on pairing between stimuli.

The finding that modularity neurons act on both sets of sensory neurons—those from the siphon as well as those from the mantle—posed an interesting question: How is the specific associative strengthening of classical conditioning achieved? Timing turned out to be an important element here. For classical conditioning to occur, the conditioned stimulus generally must precede the unconditioned stimulus by a critical and often narrow interval. For conditioning gill withdrawal by tail shock, the interval is approximately 0.5 second. If the separation is lengthened, shortened, or reversed, conditioning is drastically reduced or does not occur.

In the gill-withdrawal reflex the specificity in timing results in part from a convergence of the conditioned and unconditioned stimuli within individual sensory neurons. The unconditioned stimulus is represented in the sensory neurons by the action of the modulatory neurons, in particular the cells in which serotonin is the transmitter. The conditioned stimulus is represented by activity within the sensory neurons themselves. We found that the modulatory neurons activated by the unconditioned stimulus to the tail produce greater presynaptic facilitation of the sensory neurons if the sensory neurons had just fired action potentials in response to the conditioned stimulus. Action potentials in the sensory neurons that occur just after the tail shock have no effect.

This novel property of presynaptic facilitation is called activity dependence. Activity-dependent facilitation requires the same timing on the cellular level as does conditioning on the behavioral level and may account for such conditioning. These results suggest that a cellular mechanism of classical conditioning of the withdrawal reflex is an elaboration of presynaptic facilitation, a mechanism used for sensitization of the reflex. These experiments provided an initial suggestion that there might be a cellular alphabet for learning whereby the mechanisms of more complex types of learning may be elaborations or combinations of the mechanisms of simpler types of learning.

The next piece in the puzzle of how classical conditioning occurs was to discover why the firing of action potentials in the sensory neurons just before the unconditioned tail stimulus would enhance presynaptic facilitation. We had previously found that when serotonin is released by the modulatory neurons in response to tail shock, it initiates a series of biochemical changes in the sensory neurons [*see illustration on page 147*]. Serotonin binds to a receptor that activates an enzyme called adenylyl cyclase. This enzyme in turn converts ATP, one of the molecules that provides the energy needed to power the various activities of the cell, into cyclic AMP. Cyclic AMP then acts as a second messenger (serotonin is the first messenger) inside the cell to activate another enzyme, a protein kinase. Kinases are proteins that phosphorylate (add a phosphate group to) other proteins, thereby increasing the activity of some and decreasing the activity of others.

The activation of the protein kinase in sensory neurons has several important short-term consequences. The protein kinase phosphorylates potassium channel proteins. Phosphorylation of these channels (or of proteins that act on these channels) reduces a component of the potassium current that normally repolarizes the action potential. Reduction of potassium current prolongs the action potential and thereby allows calcium channels to be activated for longer periods, permitting more calcium to enter the presynaptic terminal. Calcium has several actions within the cell, one of which is the release of transmitter vesicles from the terminal. When, as a result of an increase in the duration of the action potentials, more calcium enters the terminal, more transmitter is released. Second, as a result of protein kinase activity, serotonin acts to mobilize transmitter vesicles from a storage pool to the release sites at the membrane; this facilitates the release of transmitter independent of an increase in calcium influx. In this action, cyclic AMP acts in parallel with another second messenger, protein kinase C, which is also activated by serotonin.

Why should the firing of action potentials in the sensory neurons just before the unconditioned stimulus enhance the action of serotonin? Action potentials produce a number of changes in the sensory neurons. They allow sodium and calcium to move in and potassium to move out, and they change the membrane potential. Abrams and Kandel found that the critical function of the action potential for activity dependence was the movement of calcium into the sensory neurons. Once in the cell, calcium binds to a protein called calmodulin, which amplifies the activation of the enzyme adenylyl cyclase by serotonin. When calcium/calmodulin binds to the adenylyl cyclase, the enzyme generates more cyclic AMP. This capacity makes adenylyl cyclase an important convergence site for the conditioned and the unconditioned stimuli.

Thus, the conditioned and the unconditioned stimuli are represented within the cell by the convergence of two different signals (calcium and serotonin) on the

CLASSICAL CONDITIONING IN *APLYSIA*

The marine snail *Aplysia* (*top left*) is used in studies of the biological basis of learning because its simple nervous system consists of only 20,000 relatively large neurons. The diagram (*bottom left*) traces one of the pathways involved in classical conditioning of the gill-withdrawal reflex in *Aplysia*. An increase in the release of neurotransmitter due to activity-dependent facilitation is a mechanism that contributes to conditioning. The molecular steps in activity-dependent facilitation are shown in the enlargement at the right. Serotonin released from the modulatory neuron by the unconditioned stimulus activates adenylyl cyclase in the sensory neuron. When the sensory neuron is active, levels of calcium are elevated within the cell. The calcium binds to calmodulin, which in turn binds to adenylyl cyclase, enhancing its ability to synthesize cyclic AMP. The cyclic AMP activates protein kinase, which leads to the release of a substantially greater amount of transmitter than would occur normally.

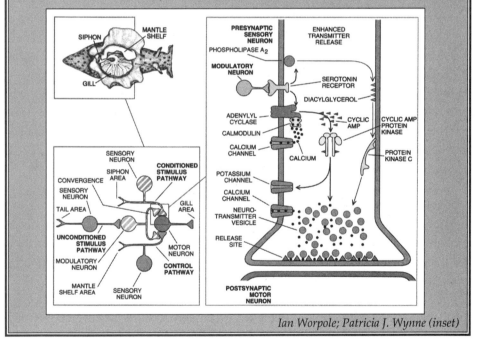

Ian Worpole; Patricia J. Wynne (inset)

same enzyme. The 0.5-second interval between the two stimuli essential for learning in the gill-withdrawal reflex may correspond to the time during which calcium is elevated in the presynaptic terminal and binds to calmodulin so as to prime the adenylyl cyclase to produce more cyclic AMP in response to serotonin.

Activity-dependent amplification of the cyclic AMP pathway is not unique to the gill- or tail-withdrawal reflexes of *Aplysia*. Genetic studies in the fruit fly *Drosophila* have implicated a similar molecular mechanism for conditioning. *Drosophila* can be conditioned, and single-gene mutants have been discovered that are deficient in learning. One such mutant, called *rutabaga,* has been studied by William G. Quinn of the Massachusetts Institute of Technology and Margaret Livingstone of Harvard University and by Yadin Dudai of the Weizmann Institute in Israel. The gene encoding the defective protein in this mutant has now been shown to be a calcium/calmodulin-dependent adenylyl cyclase. As a result of the mutation in *rutabaga,* the cyclase has lost its ability to be stimulated by calcium/calmodulin. Moreover, Ronald L. Davis and his colleagues at Cold Spring Harbor Laboratory have found that this form of the adenylyl cyclase is enriched in the mushroom bodies, a part of the fly brain critical for several types of associative learning. Thus, both cell biological studies in *Aplysia* and genetic studies in *Drosophila* point to the significance of the cyclic AMP second-messenger system in certain elementary types of implicit learning and memory storage.

What about explicit forms of learning? Do these more complex types of associative learning also have cellular representations of associativity? If so, they must differ from the mechanism of implicit learning, because, unlike classical conditioning, explicit learning is often most successful when the two events that are associated occur simultaneously. For example, we recognize the face of an acquaintance most easily when we see that acquaintance in a specific context. The stimuli of the face and of the setting act simultaneously to help us recognize the person.

As we have seen, explicit learning in humans requires the temporal lobe. Yet it was unclear at first how extensive the bilateral lesion in the temporal lobe had to be to interfere with memory storage. Subsequent studies in humans and in experimental animals by Mortimer Mishkin of the National Institutes of Health and by Squire, David G. Amaral, and Stuart Zola-Morgan of the University of California at San Diego help to answer the question. They suggest that one structure within the temporal lobe particularly critical for memory storage is the hippocampus. And yet lesions of the hippocampus interfere only with the storage of new memories: patients like H.M. still have a reasonably good memory of earlier events. The hippocampus appears to be only a temporary depository for long-term memory. The hippocampus processes the newly learned information for a period of weeks to months and then transfers the information to relevant areas of the cerebral cortex for more permanent storage. As discussed by Patricia S. Goldman-Rakic, the memory stored at these different cortical sites is then expressed through the working memory of the prefrontal cortex.

In 1973 Timothy Bliss and Terje Lømo, working in Per Andersen's laboratory in Oslo, Norway, first demonstrated that neurons in the hippocampus have remarkable plastic capabilities of the kind that would be required for learning. They found

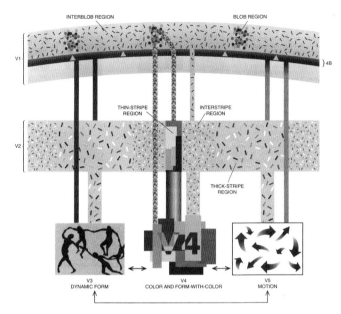

INTERBLOB REGION BLOB REGION

V1

THIN-STRIPE REGION INTERSTRIPE REGION

V2

THICK-STRIPE REGION

V4

V3
DYNAMIC FORM

V4
COLOR AND FORM-WITH-COLOR

V5
MOTION

Four perceptual pathways within the visual cortex have been identified. Color is seen when wavelength-selective cells in the blob regions of V1 send signals to specialized area V4 and also to the thin stripes of V2, which connect with V4. Form in association with color depends on connections between the interblobs of V1, the interstripes of V2 and area V4. Cells in layer 4B of V1 send signals to specialized areas V3 and V5 directly and also through the thick stripes of V2; these connections give rise to the perception of motion and dynamic form.

Damage to specialized regions of the cortex can cause strange types of blindness in which patients lose the ability to see just one attribute of the visual world, such as color, form, or motion. Artwork produced by some of these patients offers glimpses into their view of the world, as well as into the working of the visual cortex itself.

a A patient with damage to the color pathways in the cortex lost all color vision. In his drawings, a banana, a tomato, a grapefruit, and green leaves all have similar colors. (Courtesy Semir Zeki, University College London)

b When an achromatopsic patient was shown a Land color Mondrian (left) and asked to reproduce it, he was able to copy the shapes in the painting successfully. The colors within the blocks eluded him (right). (Courtesy Semir Zeki, University College London)

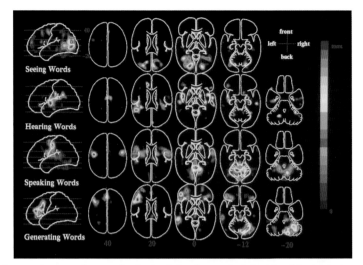

Pet scans show active neural areas. In the far left column the left side of the brain is presented; the next column shows five horizontal layers (the right side faces to the right, with the front to the top). Each row corresponds to the difference between a specific task and the control state of gazing at a dot on a television monitor. When subjects passively view nouns (row 1) the primary visual cortex lights up. When nouns are heard (row 2), the temporal lobes take command. Spoken nouns minus viewed or heard nouns (row 3) reveal motor areas used for speech. Generating verbs (row 4) requires additional neural zones, including those in the left frontal and temporal lobes corresponding roughly to Broca's and Wernicke's areas. (Courtesy Marcus Raichle, Washington University School of Medicine).

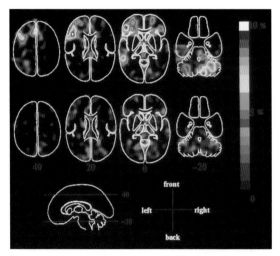

Learning-induced changes in neural activity are revealed by PET imaging. The top row shows the brain of a subject who must quickly generate verbs appropriate to visually presented nouns. The bottom row shows the result of 15 minutes of practice; the regions activated are similar to those used in simply reading out loud. (Courtesy Marcus Raichle, Washington University School of Medicine).

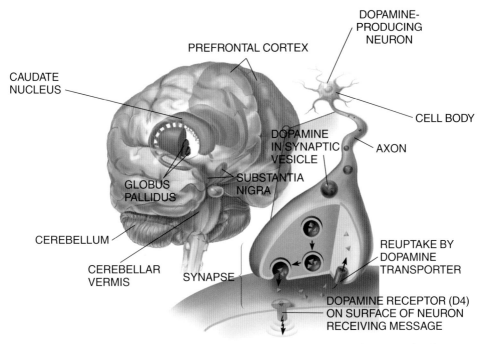

Brain structures affected in ADHD used dopamine to communicate with one another (green arrows). Genetic studies suggest that people with ADHD might have alterations in genes encoding either the D4 dopamine receptor, which receives incoming signals, or the dopamine transporter, which scavenges released dopamine for reuse. The substantia nigra, where the death of dopamine-producing neurons causes Parkinson's disease, is not affected in ADHD. (Terese Winslow)

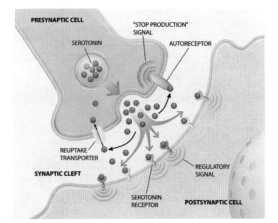

Serotonin (red spheres) secreted by a presynaptic cell binds to receptors (shades of green) on a postsynaptic cell and directs the postsynaptic cell to fire or stop firing. The cell's response is influenced by the amount of serotonin in the cleft and by the types of receptors; serotonin receptors come in at least 13 "flavors." Serotonin levels in synapses are reduced by two kinds of presynaptic molecules: autoreceptors (orange), which direct the cells to inhibit serotonin production, and reuptake transporters (yellow), which absorb the neurotransmitter. Several antidepressants, including Prozac and Paxil, increase synaptic serotonin by inhibiting its reuptake. (Tomo Narashima)

Optical illustration devised by Vilayanur S. Ramachandran illustrates the brain's ability to fill in, or construct, visual information that is missing because it falls on the blind spot of the eye. When you look at the patterns of broken green bars, the visual system produces two illusory contours defining a vertical strip. Now shut your right eye and focus on the white square in the green series of bars. Move the page toward your eye until the blue dot disappears (roughly six inches in front of your nose). Most observers report seeing the vertical strip completed across the blind spot, not the broken line. Try the same experiment with the series of just three red bars. The illusory vertical contours are less well defined, and the visual system tends to fill the horizontal bar across the blind spot. Thus, the brain fills in differently depending on the overall context of the image. (Johnny Johnson)

that a brief high-frequency train of action potentials in one of the neural pathways within the hippocampus produces an increase in synaptic strength in that pathway. The increase can be shown to last for hours in an anesthetized animal and for days and even weeks in an alert, freely moving animal.

Bliss and Lømo called this strengthening long-term potentiation (LTP). Later studies showed that LTP has different properties in different types of synapses within the hippocampus. We will focus here on an associative type of potentiation that has two interrelated characteristics. First, the associativity is of the Hebbian pre-post form: for facilitation to occur, the contributing presynaptic and postsynaptic neurons need to be active simultaneously. Second, and as a result, the long-term potentiation shows specificity: it is restricted in its action to the pathway that is stimulated.

Why is simultaneous firing of the presynaptic and postsynaptic cells necessary for long-term potentiation? The major neural pathways in the hippocampus use the amino acid glutamate as their transmitter. Glutamate produces LTP by binding to glutamate receptors on its target cells. It turns out that there are two relevant kinds of glutamate receptors: the NMDA receptors (named after the chemical N-methyl D-aspartate) and the non-NMDA receptors. Non-NMDA receptors dominate most synaptic transmission because the ion channel associated with the NMDA receptor is usually blocked by magnesium. It becomes unlocked only when the postsynaptic cell is depolarized. Moreover, optimal activation of the NMDA receptor channel requires that the two signals—glutamate binding to the receptor and depolarization of the postsynaptic cell—take place simultaneously. Thus the NMDA receptor has associative or coincidence-detecting properties much as does the adenylyl cyclase. But its temporal characteristics, a requirement for simultaneous activation, are better suited for explicit rather than implicit forms of learning.

Calcium influx into the postsynaptic cell through the unblocked NMDA receptor channel is critical for long-term potentiation, as was first shown by Gary Lynch of the University of California at Irvine and by Roger A. Nicoll and Robert S. Zucker and colleagues at the University of California at San Francisco. Calcium initiates LTP by activating at least three different types of protein kinases.

The *induction* of LTP appears to depend on postsynaptic depolarization, leading to the influx of calcium and the subsequent activation of second-messenger kinases. For the *maintenance* of LTP, on the other hand, several groups of researchers have found that enhancement of transmitter from the presynaptic terminal is involved. These workers include Bliss and his colleagues, John Bekkers and Charles Stevens of the Salk Institute, and Roberto Malinow and Richard Tsien of Stanford University.

If the induction of LTP requires a postsynaptic event (calcium influx through the NMDA receptor channels) and maintenance of LTP involves a presynaptic event

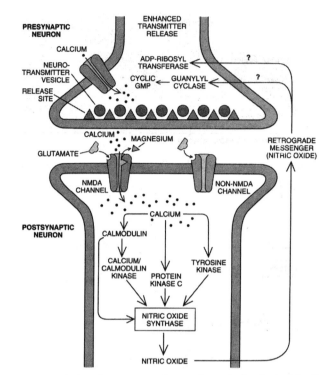

In long-term potentiation the postsynaptic membrane is depolarized by the actions of the non-NMDA receptor channels. The depolarization relieves the magnesium blockade of the NMDA channel, allowing calcium to flow through the channel. The calcium triggers calcium-dependent kinases that lead to the induction of LTP. The postsynaptic cell is thought to release a retrograde messenger capable of penetrating the membrane of the presynaptic cell. This messenger, which may be nitric oxide, is believed to act in the presynaptic terminal to enhance transmitter (glutamate) release, perhaps by activating guanylyl cyclase or ADP-ribosyl transferase. (Ian Worpole)

(increase in transmitter release), then, as first proposed by Bliss, some message must be sent from the postsynaptic to the presynaptic neurons—and that poses a problem for neuroscientists. Ever since the great Spanish anatomist Santiago Ramón y Cajal first enunciated the principle of dynamic polarization, every chemical synapse studied has proved to be unidirectional. Information flows only from the presynaptic to the postsynaptic cell. In long-term potentiation, a new principle of nerve cell communication seems to be emerging. The calcium-activated second-messenger pathways, or perhaps calcium acting directly, seem to cause release of a retrograde plasticity factor from the active postsynaptic cell. This retrograde factor then diffuses to the presynaptic terminals to activate one or more second messengers that enhance transmitter release and thereby maintain LTP [*see illustration above*].

Unlike the presynaptic terminals, which store transmitter in vesicles and release it at specialized release sites, the postsynaptic terminals lack many special release machinery. It therefore seemed attractive to posit that the retrograde messenger may be a substance that rapidly diffuses out of the postsynaptic cell across the synaptic cleft and into the presynaptic terminal. By 1991 four groups of researchers had obtained evidence that nitric oxide may be such a retrograde messenger: Thomas J. O'Dell and Ottavio Arancio in our laboratory, Erin M. Schuman and Daniel Madison of Stanford University, Paul F. Chapman and his colleagues at the University of Minnesota School of Medicine, and Georg Böhme and his colleagues in France. Inhibiting the synthesis of nitric oxide in the postsynaptic neuron or absorbing nitric oxide in the extracellular space blocks the induction of LTP, whereas applying nitric oxide enhances transmitter release from presynaptic neurons.

In the course of studying the effects of applying nitric oxide in slices of hippocampus, we and Scott A. Small and Min Zhuo made a surprising finding: we discovered that nitric oxide produces LTP only if it is paired with activity in the presynaptic neurons, much as is the case in activity-dependent presynaptic felicitation in *Aplysia*. Presynaptic activity, and perhaps calcium influx, appears to be critical for nitric oxide to produce potentiation. These experiments suggest that long-term potentiation uses a combination of two independent, associative, synaptic learning mechanisms: a Hebbian NMDA receptor mechanism and a non-Hebbian, activity-dependent, presynaptic facilitating mechanism. According to this hypothesis, the activation of NMDA receptors in the postsynaptic cells produces a retrograde signal (nitric oxide). The signal then initiates an activity-dependent presynaptic mechanism, which facilitates the release of transmitter from the presynaptic terminals.

What might be the functional advantage of combining two associative cellular mechanisms, the postsynaptic NMDA receptor and the activity-dependent presynaptic facilitation, in this way? If presynaptic facilitation is produced by a diffusible substance, that substance could, in theory, find its way into neighboring pathways. In fact, studies by Tobias Bonhoeffer and his colleagues at the Max Planck Institute for Brain Research in Frankfurt indicate that LTP initiated in one postsynaptic cell spreads to neighboring postsynaptic cells. Activity dependence of presynaptic facilitation could be a way of ensuring that only specific presynaptic pathways—those that are active—are potentiated. Any inactive presynaptic terminals would not be affected [*see illustration on page 152*].

The changes in synapses that are thought to contribute to these instances of implicit and explicit learning raise a surprising reductionist possibility. The fact that associative synaptic changes do not require complex neural networks suggests there may be a direct correspondence between these associative forms of learning and basic cellular properties. In the cases that we have reviewed the cellular prop-

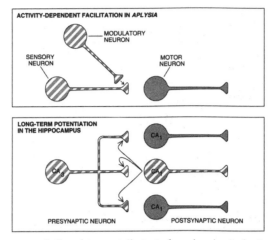

Associative processes believed to contribute to learning in Aplysia *and in the hippocampus of mammals may share similar mechanisms. Both may involve a modulatory substance that produces activity-dependent enhancement of transmitter release from the presynaptic neuron. Stripes denote neurons in which coincident activity must occur to produce the associative change. (Ian Worpole)*

erties seem to derive in turn from the properties of specific proteins—the adenylyl cyclase and the NMDA receptor—that are capable of responding to two independent signals, such as those from the conditioned stimulus and the unconditioned stimulus. Of course, these molecular associative mechanisms do not act in isolation They are embedded in cells that have rich molecular machinery for elaborating the associative process. And the cells, in turn, are embedded in the complex neural networks with considerable redundancy, parallelism, and computational power, adding substantial complexity to these elementary mechanisms.

The finding that LTP occurs in the hippocampus, a region known to be significant in memory storage, made researchers wonder whether LTP is involved in the process of storing memories in this area of the brain. Evidence that it is has been provided by Richard Morris and his colleagues at the University of Edinburgh Medical School by means of a spatial memory task. When NMDA receptors in the hippocampus are blocked, the experimental animals fail to learn the task. These experiments suggest that NMDA receptor mechanisms in the hippocampus, and perhaps LTP, are involved in spatial learning.

Having now considered the mechanisms through which learning can produce changes in nerve cells, we are faced with a final set of questions. What are the mechanisms whereby the synaptic changes produced by explicit and implicit learning endure? How is memory maintained in the long term?

Experiments in both *Aplysia* and mammals indicate that explicit and implicit memory storage proceed in stages. Storage of the initial information, a type of short-term memory, lasts minutes to hours and involves changes in the strength of existing synaptic connections (by means of second-messenger-mediated modifications of the kind we have discussed). The long-term changes (those that persist for weeks and months) are stored at the same site, but they require something entirely new: the activation of genes, the expression of new proteins, and the growth of new connections. In *Aplysia,* Craig H. Bailey, Mary C. Chen and Samuel M. Schacher and their colleagues at Columbia University, and Byrne and his colleagues at the University of Texas Health Science Center have found that stimuli that produce long-term memory for sensitization and classical conditioning lead to an increase in the number of presynaptic terminals. Similar anatomic changes occur in the hippocampus after LTP.

If long-term memory leads to anatomic changes, does that imply that our brains are constantly changing anatomically as we learn and as we forget? Will we experience changes in our brain's anatomy as a result of reading and remembering this article?

This question has been addressed by many investigators, perhaps most dramatically by Michael Merzenich of the University of California at San Francisco. Merzenich examined the representation of the hand in the sensory area of the cerebral cortex. Until recently, neuroscientists believed this representation was stable throughout life. But Merzenich and his colleagues have now demonstrated that cortical maps are subject to constant modification based on use of the sensory pathways. Since all of us are brought up in somewhat different environments, are exposed to different combinations of stimuli, and are likely to exercise our sensory and motor skills in different ways, the architecture of each of our brains will be modified in slightly different ways. This distinctive modification of brain architecture, along with a unique genetic makeup, contributes to the biological basis for the expression of individuality.

This view is best demonstrated in a study by Merzenich, in which he encouraged a monkey to touch a rotating disk with only the three middle fingers of its hand. After several thousand disk rotations, the area in the cortex devoted to the three middle fingers was expanded at the expense of that devoted to the other fingers. Practice, therefore, can lead to changes in the cortical representation of the most active fingers. What mechanisms underlie the changes? Recent evidence indicates that the cortical connections in the somatosensory system are constantly being modified and updated on the basis of correlated activity, using a mechanism that appears similar to that which generates LTP.

Indeed, early results from cell biological studies of development suggest that the mechanisms of learning may carry with them an additional bonus. There is now reason to believe that the fine-tuning of connections during late stages of development may require an activity-dependent associative synaptic mechanism perhaps similar to LTP. If that is also true on the molecular level—if learning shares common molecular mechanisms with aspects of development and growth—the study of learning may help connect cognitive psychology to the molecular biology of the organism more generally. This broad biological unification would accelerate the demystification of mental processes and position their study squarely within the evolutionary framework of biology.

—September 1992

Behavior

Sex Differences
in the Brain

Cognitive variations between the sexes reflect
differing hormonal influences on brain development.
Understanding these differences and their causes
can yield insights into brain organization

Doreen Kimura

Women and men differ not only in physical attributes and reproductive function but also in the way in which they solve intellectual problems. It has been fashionable to insist that these differences are minimal, the consequence of variations in experience during development. The bulk of the evidence suggests, however, that the effects of sex hormones on brain organization occur so early in life that from the start the environment is acting on differently wired brains in girls and boys. Such differences make it almost impossible to evaluate the effects of experience independent of physiological predisposition.

Behavioral, neurological, and endocrinologic studies have elucidated the processes giving rise to sex differences in the brain. As a result, aspects of the physiological basis for these variations have in recent years become clearer. In addition, studies of the effects of hormones on brain function throughout life suggest that the evolutionary pressures directing differences nevertheless allow for a degree of flexibility in cognitive ability between the sexes.

Major sex differences in intellectual function seem to lie in patterns of ability rather than in overall level of intelligence (IQ). We are all aware that people have different intellectual strengths. Some are especially good with words, others at using objects—for instance, at constructing or fixing things. In the same fashion, two individuals may have the same overall intelligence but have varying patterns of ability.

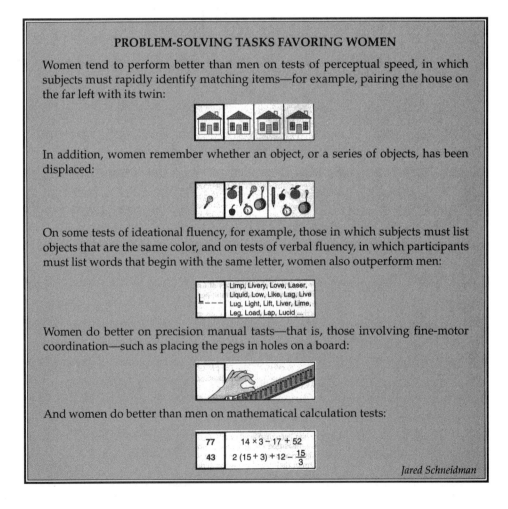

PROBLEM-SOLVING TASKS FAVORING WOMEN

Women tend to perform better than men on tests of perceptual speed, in which subjects must rapidly identify matching items—for example, pairing the house on the far left with its twin:

In addition, women remember whether an object, or a series of objects, has been displaced:

On some tests of ideational fluency, for example, those in which subjects must list objects that are the same color, and on tests of verbal fluency, in which participants must list words that begin with the same letter, women also outperform men:

L - - - Limp, Livery, Love, Laser, Liquid, Low, Like, Lag, Live Lug, Light, Lift, Liver, Lime, Leg, Load, Lap, Lucid ...

Women do better on precision manual tasts—that is, those involving fine-motor coordination—such as placing the pegs in holes on a board:

And women do better than men on mathematical calculation tests:

| 77 | $14 \times 3 - 17 + 52$ |
| 43 | $2(15+3) + 12 - \frac{15}{3}$ |

Jared Schneidman

Men, on average, perform better than women on certain spatial tasks. In particular, men have an advantage in tests that require the subject to imagine rotating an object or manipulating it in some other way. They outperform women in mathematical reasoning tests and in navigating their way through a route. Further, men are more accurate in tests of target-directed motor skills—that is, in guiding or intercepting projectiles.

Women tend to be better than men at rapidly identifying matching items, a skill called perceptual speed. They have greater verbal fluency, including the ability to find words that begin with a specific letter or fulfill some other constraint. Women also outperform men in arithmetic calculation and in recalling landmarks from a route. Moreover, women are faster at certain precision manual tasks, such as placing pegs in designated holes on a board.

Although some investigators have reported that sex differences in problem solving do not appear until after puberty, Diane Lunn, working in my laboratory at the

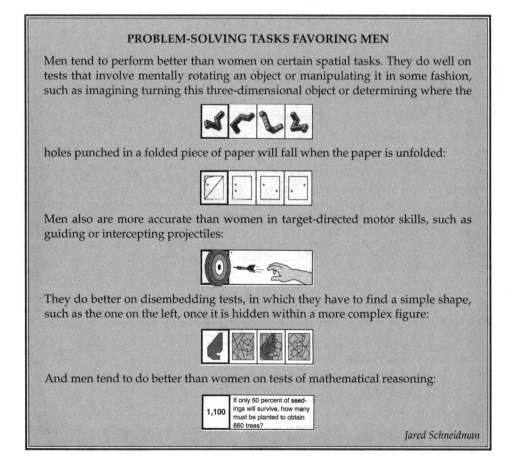

PROBLEM-SOLVING TASKS FAVORING MEN

Men tend to perform better than women on certain spatial tasks. They do well on tests that involve mentally rotating an object or manipulating it in some fashion, such as imagining turning this three-dimensional object or determining where the

holes punched in a folded piece of paper will fall when the paper is unfolded:

Men also are more accurate than women in target-directed motor skills, such as guiding or intercepting projectiles:

They do better on disembedding tests, in which they have to find a simple shape, such as the one on the left, once it is hidden within a more complex figure:

And men tend to do better than women on tests of mathematical reasoning:

1,100 | If only 60 percent of seedlings will survive, how many must be planted to obtain 660 trees?

Jared Schneidman

University of Western Ontario, and I have found three-year-old boys to be better at targeting than girls of the same age. Moreover, Neil V. Watson, when in my laboratory, showed that the extent of experience playing sports does not account for the sex difference in targeting found in young adults. Kimberly A. Kerns, working with Sheri A. Berenbaum of the University of Chicago, has found that sex differences in spatial rotation performance are present before puberty.

Differences in route learning have been systematically studied in adults in laboratory situations. For instance, Liisa Galea in my department studied undergraduates who followed a route on a tabletop map. Men learned the route in fewer trials and made fewer errors than did women. But once learning was complete, women remembered more of the landmarks than did men. These results, and those of other researchers, raise the possibility that women tend to use landmarks as a strategy to orient themselves in everyday life. The prevailing strategies used by males have not yet been clearly established, although they must relate to spatial ability.

Marion Eals and Irwin Silverman of York University studied another function that may be related to landmark memory. The researchers tested the ability of individuals to recall objects and their locations within a confined space—such as in a room or on a tabletop. Women were better able to remember whether an item had been displaced or not. In addition, in my laboratory, we measured the accuracy of object location: subjects were shown an array of objects and were later asked to replace them in their exact positions. Women did so more accurately than did men.

Imagine, for instance, that on one test the average score is 105 for women and 100 for men. If the scores for women ranged from 100 to 110 and for men from 95 to 105, the difference would be more impressive than if the women's scores ranged from 50 to 150 and the men's from 45 to 145. In the latter case, the overlap in scores would be much greater.

One measure of the variation of scores within a group is the standard deviation. To compare the magnitude of a sex difference across several distinct tasks, the difference between groups is divided by the standard deviation. The resulting number is called the effect size. Effect sizes below 0.5 are generally considered small. Based on my data, for instance, there are typically no differences between the sexes on tests of vocabulary (effect size 0.02), on nonverbal reasoning (0.03), and verbal reasoning (0.17).

On tests in which subjects match pictures, find words that begin with similar letters, or show ideational fluency—such as naming objects that are white or red—the effect sizes are somewhat larger: 0.25, 0.22, and 0.38, respectively. As discussed above, women tend to outperform men on these tasks. Researchers have reported the largest effect sizes for certain tests measuring spatial rotation (effect size 0.7) and targeting accuracy (0.75). The large effect size in these tests means there are many more men at the high end of the score distribution.

Since, with the exception of the sex chromosomes, men and women share genetic material, how do such differences come about? Differing patterns of ability between men and women most probably reflect different hormonal influences on their developing brains. Early in life the action of estrogens and androgens (male hormones chief of which is testosterone) establishes sexual differentiation. In mammals, including humans, the organism has the potential to be male or female. If a Y chromosome is present, testes or male gonads form. This development is the critical first step toward becoming a male. If the gonads do not produce male hormones or if for some reason the hormones cannot act on the tissue, the default form of the organism is female.

Once testes are formed, they produce two substances that bring about the development of a male. Testosterone causes masculinization by promoting the male, or Wolffian, set of ducts and, indirectly through conversion to dihydrotestosterone, the external appearance of scrotum and penis. The Müllerian regression factor causes the female, or Müllerian, set of ducts to regress. If anything goes wrong at any stage of the process, the individual may be incompletely masculinized.

Not only do sex hormones achieve the transformation of the genitals into male organs, but they also organize corresponding male behaviors early in life. Since we cannot manipulate the hormonal environment in humans, we owe much of what we know about the details of behavioral determination to studies in other animals. Again, the intrinsic tendency, according to studies by Robert W. Goy of the University of Wisconsin, is to develop the female pattern that occurs in the absence of masculinizing hormonal influence.

If a rodent with functional male genitals is deprived of androgens immediately after birth (either by castration or by the administration of a compound that blocks androgens), male sexual behavior, such as mounting, will be reduced. Instead female sexual behavior, such as lordosis (arching of the back), will be enhanced in adulthood. Similarly, if androgens are administered to a female directly after birth, she displays more male sexual behavior and less female behavior in adulthood.

Bruce S. McEwen and his co-workers at the Rockefeller University have shown that, in the rat, the two processes of defeminization and masculinization require somewhat different biochemical changes. These events also occur at somewhat different times. Testosterone can be converted to either estrogen (usually considered a female hormone) or dihydrotestosterone. Defeminization takes place primarily after birth in rats and is mediated by estrogen, whereas masculinization involves both dihydrotestosterone and estrogen and occurs for the most part before birth rather than after, according to studies by McEwen. A substance called alpha-fetoprotein may protect female brains from the masculinizing effects of their estrogen.

The area in the brain that organizes female and male reproductive behavior is the hypothalamus. This tiny structure at the base of the brain connects to the pituitary, the master endocrine gland. Roger A. Gorski and his colleagues at the University of California at Los Angeles have shown that a region of the preoptic area of the hypothalamus is visibly larger in male rats than in females. The size increment in males is promoted by the presence of androgens in the immediate postnatal, and to some extent prenatal, period. Laura S. Allen in Gorski's laboratory has found a similar sex difference in the human brain.

Other preliminary but intriguing studies suggest that sexual behavior may reflect further anatomic differences. In 1991 Simon LeVay of the Salk Institute for Biological Studies in San Diego reported that one of the brain regions that is usually larger in human males than in females—an interstitial nucleus of the anterior hypothalamus—is smaller in homosexual than in heterosexual men. LeVay points out that this finding supports suggestions that sexual preference has a biological substrate.

Homosexual and heterosexual men may also perform differently on cognitive tests. Brian A. Gladue of North Dakota State University and Geoff D. Sanders of City of London Polytechnic report that homosexual men perform less well on several spatial tasks than do heterosexual men. In a recent study in my laboratory, Jeff Hall found that homosexual men had lower scores on targeting tasks than did heterosexual men; however, they were superior in ideational fluency—listing things that were a particular color.

This exciting field of research is just starting, and it is crucial that investigators consider the degree to which differences in lifestyle contribute to group differences. One should also keep in mind that results concerning group differences constitute a general statistical statement; they establish a mean from which any individual may differ. Such studies are potentially a rich source of information on the physiological basis for cognitive patterns.

The lifelong effects of early exposure to sex hormones are characterized as organizational, because they appear to alter brain function permanently during a critical period. Administering the same hormones at later stages has no such effect. The hormonal effects are not limited to sexual or reproductive behaviors: they appear to extend to all known behaviors in which males and females differ. They seem to govern problem solving, aggression, and the tendency to engage in rough-and-tumble play—the boisterous body contact that young males of some mammalian species display. For example, Michael J. Meaney of McGill University finds that dihydrotestosterone, working through a structure called the amygdala rather than through the hypothalamus, gives rise to the play-fighting behavior of juvenile male rodents.

Male and female rats have also been found to solve problems differently. Christina L. Williams of Barnard College has shown that female rats have a greater tendency to use landmarks in spatial learning tasks—as it appears women do. In Williams's experiment, female rats used landmark cues, such as pictures on the wall, in preference to geometric cues, such as angles and the shape of the room. If no landmarks were available, however, females used geometric cues. In contrast, males did not use landmarks at all, preferring geometric cues almost exclusively.

Interestingly, hormonal manipulation during the critical period can alter these behaviors. Depriving newborn males of testosterone by castrating them or administering estrogen to newborn females results in a complete reversal of sex-typed behaviors in the adult animals. (As mentioned above, estrogen can have a masculinizing effect during brain development.) Treated females behave like males, and treated males behave like females.

Natural selection for reproductive advantage could account for the evolution of such navigational differences. Steven J. C. Gaulin and Randall W. FitzGerald of the University of Pittsburgh have suggested that in species of voles in which a male mates with several females rather than with just one, the range he must traverse is greater. Therefore, navigational ability seems critical to reproductive success. Indeed, Gaulin and FitzGerald found sex differences in laboratory maze learning only in voles that were polygynous, such as the meadow vole, not in monogamous species, such as the prairie vole.

Again, behavioral differences may parallel structural ones. Lucia F. Jacobs in Gaulin's laboratory has discovered that the hippocampus—a region thought to be involved in spatial learning in both birds and mammals—is larger in male polygynous voles than in females. At present, there are no data on possible sex differences in hippocampal size in human subjects.

Evidence of the influence of sex hormones on adult behavior is less direct in humans than in other animals. Researchers are instead guided by what may be parallels in other species and by spontaneously occurring exceptions to the norm in humans.

One of the most compelling areas of evidence comes from studies of girls exposed to excess androgens in the prenatal or neonatal stage. The production of abnormally large quantities of adrenal androgens can occur because of a genetic defect called congenital adrenal hyperplasia (CAH). Before the 1970s, a similar condition also unexpectedly appeared when pregnant women took various synthetic steroids. Although the consequent masculinization of the genitals can be corrected early in life and drug therapy can stop the overproduction of androgens, effects of prenatal exposure on the brain cannot be reversed.

Studies by researchers such as Anke A. Ehrhardt of Columbia University and June M. Reinisch of the Kinsey Institute have found that girls with excess exposure to androgens grow up to be more tomboyish and aggressive than their unaffected sisters. This conclusion was based sometimes on interviews with subjects and mothers, on teachers' ratings, and on questionnaires administered to the girls themselves. When ratings are used in such studies, it can be difficult to rule out the influence of expectation either on the part of an adult who knows the girls' history or on the part of the girls themselves.

Therefore, the objective observations of Berenbaum are important and convincing. She and Melissa Hines of the University of California at Los Angeles observed the play behavior of CAH-affected girls and compared it with that of their male and female siblings. Given a choice of transportation and construction toys, dolls and kitchen supplies, or books and board games, the CAH girls preferred the more typically masculine toys—for example, they played with cars for the same amount of time that normal boys did. Both the CAH girls and the boys differed from unaffected girls in their patterns of choice. Because there is every reason to think that parents would be at least as likely to encourage feminine preferences in their CAH daughters as in their unaffected daughters, these findings suggest that the toy preferences were actually altered in some way by the early hormonal environment.

Spatial abilities that are typically better in males are also enhanced in CAH girls. Susan M. Resnick, now at the National Institute on Aging, and Berenbaum and their colleagues reported that affected girls were superior to their unaffected sisters in a spatial manipulation test, two spatial rotation tests, and a disembedding test—that is, the discovery of a simple figure hidden within a more complex one. All these tasks are usually done better by males. No differences existed between the two groups on other perceptual or verbal tasks or on a reasoning task.

Studies such as these suggest that the higher the androgen levels, the better the spatial performance. But this does not seem to be the case. In 1983 Valerie J. Shute, when at the University of California at Santa Barbara, suggested that the relation between levels of androgens and some spatial capabilities might be nonlinear. In other words, spatial ability might not increase as the amount of androgen increases. Shute measured androgens in blood taken from male and female students and divided each into high- and low-androgen groups. All fell within the normal range for each sex (androgens are present in females but in very low levels). She found that in women, the high-androgen subjects were better at the spatial tests. In men the reverse was true: low-androgen men performed better.

Catherine Gouchie and I recently conducted a study along similar lines by measuring testosterone in saliva. We added tests for two other kinds of abilities: mathematical reasoning and perceptual speed. Our results on the spatial tests were very similar to Shute's: low-testosterone men were superior to high-testosterone men, but high-testosterone women surpassed low-testosterone women. Such findings suggest some optimum level of androgen for maximal spatial ability. This level may fall in the low male range.

No correlation was found between testosterone levels and performance on perceptual speed tests. On mathematical reasoning, however, the results were similar

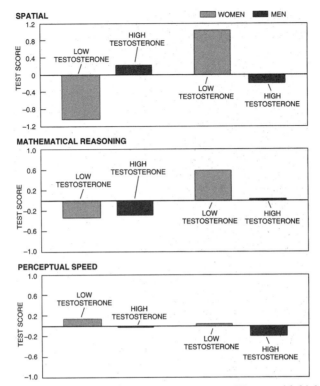

Testosterone levels can affect performance on some tests. Women with high levels of testosterone perform better on a spatial test (top) than do women with low levels; men with low levels outperform men with high levels. On a mathematical reasoning test (middle), low testosterone corresponds to better performance in men; in women there is no such relation. On a test in which women usually excel (bottom), no relation is found between testosterone and performance.

to those of spatial ability tests for men: low-androgen men tested higher, but there was no obvious relation in women.

Such findings are consistent with the suggestion by Camilla P. Benbow of Iowa State University that high mathematical ability has a significant biological determinant. Benbow and her colleagues have reported consistent sex differences in mathematical reasoning ability favoring males. These differences are especially sharp at the upper end of the distribution, where males outnumber females 13 to one. Benbow argues that these differences are not readily explained by socialization.

It is important to keep in mind that the relation between natural hormonal levels and problem solving is based on correlational data. Some form of connection between the two measures exists, but how this association is determined or what its causal basis may be is unknown. Little is currently understood about the relation between adult levels of hormones and those in early life, when abilities appear to be organized in the nervous system. We have a lot to learn about the precise mechanisms underlying cognitive patterns in people.

Another approach to probing differences between male and female brains is to examine and compare the functions of particular brain systems. One noninvasive way to accomplish this goal is to study people who have experienced damage to a specific brain region. Such studies indicate that the left half of the brain in most people is critical for speech, the right for certain perceptual and spatial functions.

It is widely assumed by many researchers studying sex differences that the two hemispheres are more asymmetrically organized for speech and spatial functions in men than in women. This idea comes from several sources. Parts of the corpus callosum, a major neural system connecting the two hemispheres, may be more extensive in women; perceptual techniques that probe brain asymmetry in normal-functioning people sometimes show smaller asymmetries in women than in men, and damage to one brain hemisphere sometimes has a lesser effect in women than the comparable injury has in men.

In 1982 Marie-Christine de Lacoste, now at the Yale University School of Medicine, and Ralph L. Holloway of Columbia University reported that the back part of the corpus callosum, an area called the splenium, was larger in women than in men. This finding has subsequently been both refuted and confirmed. Variations in the shape of the corpus callosum that may occur as an individual ages as well as different methods of measurement may produce some of the disagreements. Most recently, Allen and Gorski found the same sex-related size difference in the splenium.

The interest in the corpus callosum arises from the assumption that its size may indicate the number of fibers connecting the two hemispheres. If more connecting fibers existed in one sex, the implication would be that in that sex the hemispheres communicate more fully. Although sex hormones can alter callosal size in rats, as Victor H. Denenberg and his associates at the University of Connecticut have demonstrated, it is unclear whether the actual number of fibers differs between the sexes. Moreover, sex differences in cognitive function have yet to be related to a difference in callosal size. New ways of imaging the brain in living humans will undoubtedly increase knowledge in this respect.

The view that a male brain is functionally more asymmetric than a female brain is long-standing. Albert M. Galaburda of Beth Israel Hospital in Boston and the late Norman Geschwind of Harvard Medical School proposed that androgens increased the functional potency of the right hemisphere. In 1981 Marian C. Diamond of the University of California at Berkeley found that the right cortex is thicker than the left in male rats but not in females. Jane Stewart of Concordia University in Montreal, working with Bryan E. Kolb of the University of Lethbridge in Alberta, recently pinpointed early hormonal influences on this asymmetry: androgens appear to suppress left cortex growth.

Last year de Lacoste and her colleagues reported a similar pattern in human fetuses. They found the right cortex was thicker than the left in males. Thus, there appear to be some anatomic reasons for believing that the two hemispheres might not be equally asymmetric in men and women.

Despite this expectation, the evidence in favor of it is meager and conflicting, which suggests that the most striking sex differences in brain organization may not be related to asymmetry. For example, if overall differences between men and women in spatial ability were related to differing right hemispheric dependence for such functions, then damage to the right hemisphere would perhaps have a more devastating effect on spatial performance in men.

My laboratory has recently studied the ability of patients with damage to one hemisphere of the brain to rotate certain objects mentally. In one test, a series of line drawings of either a left or a right gloved hand is presented in various orientations. The patient indicates the hand being depicted by simply pointing to one of two stuffed gloves that are constantly present.

The second test uses two three-dimensional blocklike figures that are mirror images of one another. Both figures are present throughout the tests. The patient is given a series of photographs of these objects in various orientations, and he or she must place each picture in front of the object it depicts. (These nonverbal procedures are employed so that patients with speech disorders can be tested.)

As expected, damage to the right hemisphere resulted in lower scores for both sexes on these tests than did damage to the left hemisphere. Also as anticipated, women did less well than men on the block spatial rotation test. Surprisingly, however, damage to the right hemisphere had no greater effect in men than in women. Women were at least as affected as men by damage to the right hemisphere. This result suggests that the normal differences between men and women on such rotational tests are not the result of differential dependence on the right hemisphere. Some other brain systems must be mediating the higher performance by men.

Parallel suggestions of greater asymmetry in men regarding speech have rested on the fact that the incidence of aphasias, or speech disorders, are higher in men than in women after damage to the left hemisphere. Therefore, some researchers have found it reasonable to conclude that speech must be more bilaterally organized in women. There is, however, a problem with this conclusion. During my 20 years of experience with patients, aphasia has not been disproportionately present in women with right hemispheric damage.

In searching for an explanation, I discovered another striking difference between men and women in brain organization of speech and related motor function. Women are more likely than men to suffer aphasia when the front part of the brain is damaged. Because restricted damage within a hemisphere more frequently affects the posterior than the anterior area in both men and women, this differential

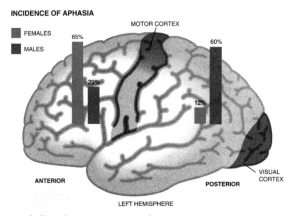

Aphasias, or speech disorders, occur most often in women when damage is to the front of the brain. In men, they occur more frequently when damage is in the posterior region. The data presented above derive from one set of patients. (Jared Schneidman)

dependence may explain why women incur aphasia less often than do men. Speech functions are thus less likely to be affected in women not because speech is more bilaterally organized in women but because the critical area is less often affected.

A similar pattern emerges in studies of the control of hand movements, which are programmed by the left hemisphere. Apraxia, or difficulty in selecting appropriate hand movements, is very common after left hemisphere damage. It is also strongly associated with difficulty in organizing speech. In fact, the critical functions that depend on the left hemisphere may relate not to language per se but to organization of the complex oral and manual movements on which human communication systems depend. Studies of patients with left hemispheric damage have revealed that such motor selection relies on anterior systems in women but on posterior systems in men.

The synaptic proximity of women's anterior motor selection system (or "praxis system") to the motor cortex directly behind it may enhance fine-motor skills. In contrast, men's motor skills appear to emphasize targeting or directing movements toward external space—some distance away from the self. There may be advantages to such motor skills when they are closely meshed with visual input to the brain, which lies in the posterior region.

Women's dependence on the anterior region is detectable even when tests involve using visual guidance—for instance, when subjects must build patterns with blocks by following a visual model. In studying such a complex task, it is possible to compare the effects of damage to the anterior and posterior regions of both hemispheres because performance is affected by damage to either hemisphere. Again, women prove more affected by damage to the anterior region of the right hemisphere than by posterior damage. Men tend to display the reverse pattern.

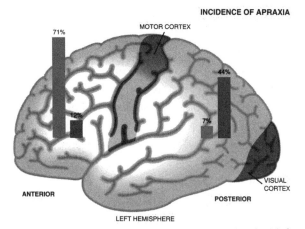

Apraxia, or difficulty in selecting hand movements, is associated with frontal damage to the left hemisphere in women and with posterior damage in men. It is also associated with difficulties in organizing speech. (Jared Schneidman)

Although I have not found evidence of sex differences in functional brain asymmetry with regard to basic speech, motor selection, or spatial rotation ability, I have found slight differences in more abstract verbal tasks. Scores on a vocabulary test, for instance, were affected by damage to either hemisphere in women, but such scores were affected only by left-sided injury in men. This finding suggests that in reviewing the meanings of words, women use the hemispheres more equally than do men.

In contrast, the incidence of nonright-handedness, which is presumably related to lesser left hemispheric dependence, is higher in men than in women. Even among right-handers, Marion Annett, now at the University of Leicester in the U.K., has reported that women are more right-handed than men—that is, they favor their right hand even more than do right-handed men. It may well be, then, that sex differences in asymmetry vary with the particular function being studied and that it is not always the same sex that is more asymmetric.

Taken altogether, the evidence suggests that men's and women's brains are organized along different lines from very early in life. During development, sex hormones direct such differentiation. Similar mechanisms probably operate to produce variation within sexes, since there is a relation between levels of certain hormones and cognitive makeup in adulthood.

One of the most intriguing findings is that cognitive patterns may remain sensitive to hormonal fluctuations throughout life. Elizabeth Hampson of the University of Western Ontario showed that the performance of women on certain tasks changed throughout the menstrual cycle as levels of estrogen went up or down. High levels of the hormone were associated not only with relatively depressed spatial ability but also with enhanced articulatory and motor capability.

In addition, I have observed seasonal fluctuations in spatial ability in men. Their performance is improved in the spring when testosterone levels are lower. Whether these intellectual fluctuations are of any adaptive significance or merely represent ripples on a stable baseline remains to be determined.

To understand human intellectual functions, including how groups may differ in such functions, we need to look beyond the demands of modern life. We did not undergo natural selection for reading or for operating computers. It seems clear that the sex differences in cognitive patterns arose because they proved evolutionarily advantageous. And their adaptive significance probably rests in the distant past. The organization of the human brain was determined over many generations by natural selection. As studies of fossil skulls have shown, our brains are essentially like those of our ancestors of 50,000 or more years ago.

For the thousands of years during which our brain characteristics evolved, humans lived in relatively small groups of hunter-gatherers. The division of labor between the sexes in such a society probably was quite marked, as it is in existing hunter-gatherer societies. Men were responsible for hunting large game, which often required long-distance travel. They were also responsible for defending the group against predators and enemies and for the shaping and use of weapons. Women most probably gathered food near the camp, tended the home, prepared food and clothing, and cared for children.

Such specializations would put different selection pressures on men and women. Men would require long-distance route-finding ability so they could recognize a geographic array from varying orientations. They would also need targeting skills. Women would require short-range navigation, perhaps using landmarks, fine-motor capabilities carried on within a circumscribed space, and perceptual discrimination sensitive to small changes in the environment or in children's appearance or behavior.

The finding of consistent and, in some cases, quite substantial sex differences suggests that men and women may have different occupational interests and capabilities, independent of societal influences. I would not expect, for example, that men and women would necessarily be equally represented in activities or professions that emphasize spatial or math skills, such as engineering or physics. But I might expect more women in medical diagnostic fields where perceptual skills are important. So that even though any one individual might have the capacity to be in a "nontypical" field, the sex proportions as a whole may vary.

—September 1992

EVIDENCE FOR
A BIOLOGICAL INFLUENCE
IN MALE HOMOSEXUALITY

*Two pieces of evidence, a structure within
the human brain and a genetic link, point
to a biological component for male homosexuality*

Simon LeVay and Dean H. Hamer

Most men are sexually attracted to women, most women to men. To many people, this seems only the natural order of things—the appropriate manifestation of biological instinct, reinforced by education, religion, and the law. Yet a significant minority of men and women—estimates range from 1 to 5 percent—are attracted exclusively to members of their own sex. Many others are drawn, in varying degrees, to both men and women.

How are we to understand such diversity in sexual orientation? Does it derive from variations in our genes or our physiology, from the intricacies of our personal history, or from some confluence of these? Is it for that matter a choice rather than a compulsion?

Probably no one factor alone can elucidate so complex and variable a trait as sexual orientation. But recent laboratory studies, including our own, indicate that genes and brain development play a significant role. How, we do not yet know. It may be that genes influence the sexual differentiation of the brain and its interaction with the outside world, thus diversifying its already vast range of responses to sexual stimuli.

The search for biological roots of sexual orientation has run along two broad lines. The first draws on observations made in yet another hunt—that for physical differences between men's and women's brains. As we shall see, "gay" and "straight" brains may be differentiated in curiously analogous fashion. The second approach is to scout out genes by studying the patterns in which homosexuality occurs in families and by directly examining the hereditary material, DNA.

Researchers have long sought within the human brain some manifestation of the most obvious classes into which we are divided—male and female. Such sex differentiation of the brain's structure, called sexual dimorphism, proved hard to establish. On average, a man's brain has a slightly larger size that goes along with his larger body; other than that, casual inspection does not reveal any obvious dissimilarity between the sexes. Even under a microscope, the architecture of men's and women's brains is very similar. Not surprisingly, the first significant observations of sexual dimorphism were made in laboratory animals.

Of particular importance is a study of rats conducted by Roger A. Gorski of the University of California at Los Angeles. In 1978 Gorski was inspecting the rat's hypothalamus, a region at the base of its brain that is involved in instinctive behaviors and the regulation of metabolism. He found that one group of cells near the front of the hypothalamus is several times larger in male than in female rats. Although this cell group is very small, less than a millimeter across even in males, the difference between the sexes is quite visible in appropriately stained slices of tissue, even without the aid of a microscope.

Gorski's finding was especially interesting because the general region of the hypothalamus in which this cell group occurs, known as the medial preoptic area, has been implicated in the generation of sexual behavior—in particular, behaviors typically displayed by males. For example, male monkeys with damaged medial preoptic areas are apparently indifferent to sex with female monkeys, and electrical stimulation of this region can make an inactive male monkey approach and mount a female. It should be said, however, that we have yet to find in monkeys a cell group analogous to the sexually dimorphic one occurring in rats.

Nor is the exact function of the rat's sexually dimorphic cell group known. What is known, from a study by Gorski and his co-workers, is that androgens—typical male hormones—play a key role in bringing about the dimorphism during development. Neurons within the cell group are rich in receptors for sex hormones, both for androgens—testosterone is the main representative—and for female hormones known as estrogens. Although male and female rats initially have about the same numbers of neurons in the medial preoptic area, a surge of testosterone secreted by the testes of male fetuses around the time of birth acts to stabilize their neuronal population. In females the lack of such a surge allows many neurons in this cell group to die, leading to the typically smaller structure. Interestingly, it is only for a few days before and after birth that the medial preoptic neurons are sensitive to androgen; removing androgens in an adult rat by castration does not cause the neurons to die.

Gorski and his colleagues at U.C.L.A., especially his student Laura S. Allen, have also found dimorphic structures in the human brain. A cell group named INAH3 (derived from "third interstitial nucleus of the anterior hypothalamus") in the medial preoptic region of the hypothalamus is about three times larger in men than in women. (Notably, however, size varies considerably even within one sex.)

In 1990 one of us (LeVay) decided to check whether INAH3 or some other cell group in the medial preoptic area varies in size with sexual orientation as well as with sex. This hypothesis was something of a long shot, given the prevailing notion that sexual orientation is a "high-level" aspect of personality molded by environment and culture. Information from such elevated sources is thought to be processed primarily by the cerebral cortex and not by "lower" centers such as the hypothalamus.

LeVay examined the hypothalamus in autopsy specimens from 19 homosexual men, all of whom had died of complications of AIDS, and 16 heterosexual men, six of whom had also died of AIDS. (The sexual orientation of those who had died of non-AIDS causes was not determined. But assuming a distribution similar to that of the general populace, no more than one or two of them were likely to have been gay.) LeVay also included specimens from six women whose sexual orientation was unknown.

After encoding the specimens to eliminate subjective bias, LeVay cut each hypothalamus into serial slices, stained these to mark the neuronal cell groups, and measured their cross-sectional areas under a microscope. Armed with information about the areas, plus the thickness of the slices, he could readily calculate the volumes of each cell group. In addition to Allen and Gorski's sexually dimorphic nucleus INAH3, LeVay examined three other nearby groups—INAH1, INAH2, and INAH4.

Like Allen and Gorski, LeVay observed that INAH3 was more than twice as large in the men as in the women. But INAH3 was also between two and three times larger in the straight men than in the gay men. In some gay men, the cell group was

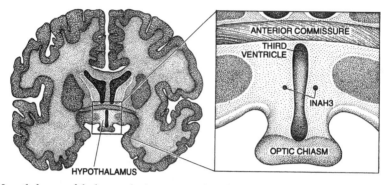

Hypothalamus of the human brain was examined for differences related to sexual orientation. The hypothalamus of each of the forty-one subjects was stained to mark neuronal cell groups. The cell group termed INAH3 in the medial preoptic area was more than twice as large in the men as it was in the women. INAH3 also turned out to be two to three times larger in straight men than it was in gay men. This finding suggests a difference related to male sexual orientation about as great as that related to sex. (Patricia J. Wynne)

altogether absent. Statistical analysis indicated that the probability of this result's being attributed to chance was about one in 1,000. In fact, there was no significant difference between volumes of INAH3 in the gay men and in the women. So the investigation suggested a dimorphism related to male sexual orientation about as great as that related to sex.

A primary concern in such a study is whether the observed structural differences are caused by some variable other than the one of interest. A major suspect here was AIDS. The AIDS virus itself, as well as other infectious agents that take advantage of a weakened immune system, can cause serious damage to brain cells. Was this the reason for the small size of INAH3 in the gay men, all of whom had died of AIDS?

Several lines of evidence indicate otherwise. First, the heterosexual men who died of AIDS had INAH3 volumes no different from those who died of other causes. Second, the AIDS victims with small INAH3s did not have case histories distinct from those with large INAH3s; for instance, they had not been ill longer before they died. Third, the other three cell groups in the medial preoptic area—INAH1, INAH2, and INAH4—turned out to be no smaller in the AIDS victims. If the disease were having a nonspecific destructive effect, one would have suspected otherwise. Finally, after completing the main study, LeVay obtained the hypothalamus of one gay man who had died of non-AIDS causes. This specimen, processed "blind" along with several specimens from heterosexual men of similar age, confirmed the main study: the volume of INAH3 in the gay man was less than half that of INAH3 in the heterosexual men.

One other feature in brains that is related to sexual orientation has been reported by Allen and Gorski. They found that the anterior commissure, a bundle of fibers running across the midline of the brain, is smallest in heterosexual men, larger in women, and largest in gay men. After correcting for overall brain size, the anterior commissure in women and in gay men were comparable in size.

What might be behind these apparent correlations between sexual orientation and brain structure? Logically, three possibilities exist. One is that the structural differences were present early in life—perhaps even before birth—and helped to establish the men's sexual orientation. The second is that the differences arose in adult life as a result of the men's sexual feelings or behavior. The third possibility is that there is no causal connection, but both sexual orientation and the brain structures in question are linked to some third variable, such as a developmental event during uterine or early postnatal life.

We cannot decide among these possibilities with any certainty. On the basis of animal research, however, we find the second scenario, that the structural differences came about in adulthood, unlikely. In rats, for example, the sexually dimorphic cell group in the medial preoptic area appears plastic in its response to androgens during early brain development but later is largely resistant to change. We favor the first possibility, that the structural differences arose during the period of brain development and consequently contributed to sexual behavior. Because the medial preoptic region of the hypothalamus is implicated in sexual behavior in monkeys, the size of INAH3 in men may indeed influence sexual orientation. But such a causal connection is speculative at this point.

Assuming that some of the structural differences related to sexual orientation were present at birth in certain individuals, how did they arise? One candidate is the interaction between gonadal steroids and the developing brain; this interaction is responsible for differences in the structure of male and female brains. A number of scientists have speculated that atypical levels of circulating androgens in some fetuses cause them to grow into homosexual adults. Specifically, they suggest that androgen levels are unusually low in male fetuses that become gay and unusually high in female fetuses that become lesbian.

A more likely possibility is that there are intrinsic differences in the way individual brains respond to androgens during development, even when the hormone levels are themselves no different. This response requires a complex molecular machinery, starting with the androgen receptors but presumably including a variety of proteins and genes whose identity and roles are still unknown.

At first glance, the very notion of gay genes might seem absurd. How could genes that draw men or women to members of the same sex survive the Darwinian screening for reproductive fitness? Surely the parents of most gay men and lesbians are heterosexual? In view of such apparent incongruities, research focuses on genes that sway rather than determine sexual orientation. The two main approaches to seeking such genes are twin and family studies and DNA linkage analysis.

Twin and family tree studies are based on the principle that genetically influenced traits run in families. The first modern study on the patterns of homosexuality within families was published in 1985 by Richard C. Pillard and James D. Weinrich of Boston University. Since then, five other systematic studies of the twins and siblings of gay men and lesbians have been reported.

The pooled data for men show that about 57 percent of identical twins, 24 percent of fraternal twins, and 13 percent of brothers of gay men are also gay. For women, approximately 50 percent of identical twins, 16 percent of fraternal twins, and 13 percent of sisters of lesbians are also lesbian. When these data are compared with baseline rates of homosexuality, a good amount of family clustering of sexual orientation becomes evident for both sexes. In fact, J. Michael Bailey of Northwestern University and his co-workers estimate that the overall heritability of sexual orientation—that proportion of the variance in a trait that comes from genes—is about 53 percent for men and 52 percent for women. (The family clustering is most obvious for relatives of the same sex, less so for male-female pairs.)

To evaluate the genetic component of sexual orientation and to clarify its mode of inheritance, we need a systematic survey of the extended families of gay men and lesbians. One of us (Hamer), Stella Hu, Victoria L. Magnuson, Nan Hu, and Angela M. L. Pattatucci of the National Institutes of Health have initiated such a study. It is part of a larger one by the National Cancer Institute to investigate risk factors for certain cancers that are more frequent in some segments of the gay population.

Hamer and his colleagues' initial survey of males confirmed the sibling results of Pillard and Weinrich. A brother of a gay man had a 14 percent likelihood of being gay as compared with 2 percent for the men without gay brothers. (The study used an unusually stringent definition of homosexuality, leading to the low average rate.) Among more distant relatives, an unexpected pattern showed up: maternal uncles had a 7 percent chance of being gay, whereas sons of maternal aunts had an 8 percent chance. Fathers, paternal uncles, and the three other types of cousins showed no correlation at all.

Although this study pointed to a genetic component, homosexuality occurred much less frequently than a single gene inherited in simple Mendelian fashion would suggest. One interpretation, that genes are more important in some families than in others, is borne out by looking at families having two gay brothers. Compared with randomly chosen families, rates of homosexuality in maternal uncles increased from 7 to 10 percent and in maternal cousins from 8 to 13 percent. This familial clustering, even in relatives outside the nuclear family, presents an additional argument for a genetic root to sexual orientation.

Why are most gay male relatives of gay men on the mother's side of the family? One possibility—that the subjects somehow knew more about their maternal relatives—seems unlikely because opposite-sex gay relatives of gay males and lesbians were equally distributed between both sides of the family. Another explanation is that homosexuality, while being transmitted by both parents, is expressed only in one sex—in this case, males. When expressed, the trait reduces the reproductive rate and must therefore be disproportionately passed on by the mother. Such an effect may partially account for the concentration of gay men's gay relatives on the maternal side of the family. But proof of this hypothesis will require finding an appropriate gene on an autosomal chromosome, which is inherited from either parent.

A third possibility is X chromosome linkage. A man has two sex chromosomes: a Y, inherited from his father, and an X, cut and pasted from the two X chromosomes carried by his mother. Therefore, any trait that is influenced by a gene on the X chromosome will tend to be inherited through the mother's side and will be preferentially observed in brothers, maternal uncles, and maternal cousins, which is exactly the observed pattern.

To test this hypothesis, Hamer and his colleagues embarked on a linkage study of the X chromosome in gay men. Linkage analysis is based on two principles of genetics. If a trait is genetically influenced, then relatives who share the trait will share the gene more often than is expected by chance—this is true even if the gene plays only a small part. Also, genes that are close together on a chromosome are almost always inherited together. Therefore, if there is a gene that influences sexual orientation, it should be "linked" to a nearby DNA marker that tends to travel along with it in families. For traits affected by only one gene, linkage can precisely locate the gene on a chromosome. But for complex traits such as sexual orientation, linkage also helps to determine whether a genetic component really exists.

To initiate a linkage analysis of male sexual orientation, the first requirement was to find informative markers, segments of DNA that flag locations on a chromosome. Fortunately, the Human Genome Project has already generated a large catalogue of markers spanning all of the X chromosomes. The most useful ones are short, repeated DNA sequences that have slightly different lengths in different persons. To detect the markers, the researchers used the polymerase chain reaction to make several billion copies of specific regions of the chromosome and then separated the different fragments by the method of gel electrophoresis.

The second step in the linkage analysis was to locate suitable families. When scientists study simple traits such as color blindness or sickle cell anemia—which involve a single gene—they tend to analyze large, multigenerational families in which each member clearly either has or does not have the trait. Such an approach was unsuited for studying sexual orientation. First, identifying someone as not homosexual is tricky; the person may be concealing his or her true orientation or may not be aware of it. Because homosexuality was even more stigmatized in the past, multigenerational families are especially problematic in this regard. Moreover, genetic modeling shows that for traits that involve several different genes expressed at varying levels, studying large families can actually decrease the chances of finding a linked gene: too many exceptions are included.

For these reasons, Hamer and his co-workers decided to focus on nuclear families with two gay sons. One advantage of this approach is that individuals who say they are homosexual are unlikely to be mistaken. Furthermore, the approach can detect a single linked gene even if other genes or noninherited factors are required for its expression. For instance, suppose that being gay requires an X chromosome gene together with another gene on an autosome, plus some set of environmental circumstances. Studying gay brothers would give a clear-cut result because both would have the X chromosome gene. In contrast, heterosexual brothers of gay men would sometimes share the X chromosome gene and sometimes not, leading to confusing results.

Genetic analysts now believe that studying siblings is the key to traits that are affected by many elements. Because Hamer and his colleagues were most interested in finding a gene that expresses itself only in men but is transmitted through women, they restricted their search to families with gay men but no gay father-gay son pairs.

Forty such families were recruited. DNA samples were prepared from the gay brothers and, where possible, from their mothers or sisters. The samples were typed for 22 markers that span the X chromosome from the tip of the short arm to the end of the long arm. At each marker, a pair of gay brothers was scored as concordant if they inherited identical markers from their mother or as discordant if they inherited different ones. Fifty percent of the markers were expected to be identical by chance. Corrections were also made for the possibility of the mothers having two copies of the same marker.

The results of this study were striking. Over most of the X chromosome the markers were randomly distributed between the gay brothers. But at the tip of the long arm of the X chromosome, in a region known as Xq28, there was a considerable excess of concordant brothers: 33 pairs shared the same marker, whereas only seven pairs did not. Although the sample size was not large, the result was statistically significant: the probability of such a skewed ratio occurring by chance alone is less than one in 200. In a control group of 314 randomly selected pairs of brothers,

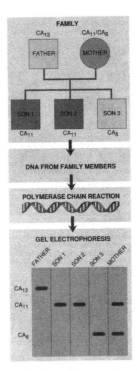

Pinpointing genes shared by gay brothers (son 1 and son 2) first involved taking DNA from subjects. Several billion copies of specific regions of the X chromosome were then made using the polymerase chain reaction and the different fragments were separated by gel electrophoresis. Gay brothers shared a marker, in this hypothetical example CA_{11} in the Xq28 region at rates far greater than predicted by chance. (Edward Bell)

most of whom can be presumed to be heterosexual, Xq28 markers were randomly distributed.

The most straightforward interpretation of the finding is that chromosomal region Xq28 contains a gene that influences male sexual orientation. The study provides the strongest evidence to date that human sexuality is influenced by heredity because it directly examines the genetic information, the DNA. But as with all initial studies, there are some caveats.

First, the result needs to be replicated: several other claims of finding genes related to personality traits have proved controversial. Second, the gene itself has not yet been isolated. The study locates it within a region of the X chromosome that is about four million base pairs in length. This region represents less than 0.2 percent of the total human genome, but it is still large enough to contain several hundred genes. Finding the needle in this haystack will require either large numbers of families or more complete information about the DNA sequence to identify all possi-

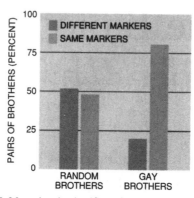

Gene sharing in the Xq28 region is significantly greater in gay brothers than in the general population. Of 40 pairs of gay brothers studied, 33 pairs shared the Xq28 region. In a control group of 314 randomly selected pairs of brothers, Xq28 markers were found to be almost equally distributed. (Edward Bell)

ble coding regions. As it happens, Xq28 is extraordinarily rich in genetic loci and will probably be one of the first regions of the human genome to be sequenced in its entirety.

A third caveat is that researchers do not know quantitatively how important a role Xq28 plays in male sexual orientation. Within the population of gay brothers studied, seven of 40 brothers did not share markers. Assuming that 20 siblings should inherit identical markers by chance alone, 36 percent of the gay brothers show no link between homosexuality and Xq28. Perhaps these men inherited different genes or were influenced by nongenetic physiological factors or by the environment. Among all gay men—most of whom do not have gay brothers—the influence of Xq28 is even less clear. Also unknown is the role of Xq28, and other genetic loci, in female sexual orientation.

How might a genetic locus at Xq28 affect sexuality? One idea is that the hypothetical gene affects hormone synthesis or metabolism. A candidate for such a gene was the androgen receptor locus, which encodes a protein essential for masculinization of the human brain and is, moreover, located on the X chromosome. To test this idea, Jeremy Nathans, Jennifer P. Macke, Van L. King, and Terry R. Brown of Johns Hopkins University teamed up with Bailey of Northwestern and Hamer, Hu, and Hu of the NIH. They compared the molecular structure of the androgen receptor gene in 197 homosexual men and 213 predominantly heterosexual men. But no significant variations in the protein coding sequences were found. Also, linkage studies showed no correlation between homosexuality in brothers and inheritance of the androgen receptor locus. Most significant of all, the locus turned out to be at

Xq11, far from the Xq28 region. This study excludes the androgen receptor from playing a significant role in male sexual orientation.

A second idea is that the hypothetical gene acts indirectly, through personality or temperament, rather than directly on sexual-object choice. For example, people who are genetically self-reliant might be more likely to acknowledge and act on same-sex feelings than are people who are dependent on the approval of others.

Finally, the intriguing possibility arises that the Xq28 gene product bears directly on the development of sexually dimorphic brain regions such as INAH3. At the simplest level, such an agent could act autonomously, perhaps in the womb, by stimulating the survival of specific neurons in preheterosexual males or by promoting their death in females and prehomosexual men. In a more complex model, the gene product could change the sensitivity of a neuronal circuit in the hypothalamus to stimulation by environmental cues—perhaps in the first few years of life. Here the genes serve to predispose rather than to predetermine. Whether this fanciful notion contains a grain of truth remains to be seen. It is in fact experimentally testable, using current tools of molecular genetics and neurobiology.

Our research has attracted an extraordinary degree of public attention, not so much because of any conceptual breakthrough—the idea that genes and the brain are involved in human behavior is hardly new—but because it touches on a deep conflict in contemporary American society. We believe scientific research can help dispel some of the myths about homosexuality that in the past have clouded the image of lesbians and gay men. We also recognize, however, that increasing knowledge of biology may eventually bring with it the power to infringe on the natural rights of individuals and to impoverish the world of its human diversity. It is important that our society expand discussions of how new scientific information should be used to benefit the human race in its entirety.

—May 1994

THE BIOLOGICAL
EVIDENCE CHALLENGED

*Even if genetic and neuroanatomical traits
turn out to be correlated with sexual orientation,
causation is far from proved*

William Byne

Human-rights activists, religious organizations, and all three branches of the U.S. government debate whether sexual orientation is biological. The discussion has grabbed headlines, but behavioral scientists find it passé. The salient question about biology and sexual orientation is not whether biology is involved but how it is involved. All psychological phenomena are ultimately biological.

Even if the public debate were more precisely framed, it would still be misguided. Most of the links in the chain of reasoning from biology to sexual orientation and social policy do not hold up under scrutiny. At the political level, a requirement that an unconventional trait be inborn or immutable is an inhumane criterion for a society to use in deciding which of its nonconformists it will grant tolerance. Even if homosexuality were entirely a matter of choice, attempts to extirpate it by social and criminal sanctions devalue basic human freedoms and diversity.

Furthermore, the notion that homosexuality must be either inborn and immutable or freely chosen is in turn misinformed. Consider the white-crowned sparrow, a bird that learns its native song during a limited period of development. Most sparrows exposed to a variety of songs, including that of their own species, will learn their species's song, but some do not. After a bird has learned a song, it can neither unlearn that song nor acquire a new one. Although sexual orientation is not a matter of mimicry, it is clear that learned behavior can nonetheless be immutable.

Finally, what evidence exists thus far of innate biological traits underlying homosexuality is flawed. Genetic studies suffer from the inevitable confounding of nature and nurture that plagues attempts to study heritability of psychological traits. Investigations of the brain rely on doubtful hypotheses about differences between the brains of men and women. Biological mechanisms that have been proposed to explain the existence of gay men often cannot be generalized to explain the existence of lesbians (whom studies have largely neglected). And the continuously graded nature of most biological variables is at odds with the paucity of adult bisexuals suggested by most surveys.

To understand how biological factors influence sexual orientation, one must first define orientation. Many researchers, most conspicuously Simon LeVay, treat it as a sexually dimorphic trait: men are generally "programmed" for attraction to women, and women are generally programmed for attraction to men. Male homosexuals, according to this framework, have female programming, and lesbians have male programming. Some researchers suggest that this programming is accomplished by biological agents, perhaps even before birth; others believe it occurs after birth in response to social factors and subjective experiences. As the function of the brain is undoubtedly linked to its structure and physiology, it follows that homosexuals' brains might exhibit some features typical of the opposite sex.

The validity of this "intersex" expectation is questionable. For one, sexual orientation is not dimorphic; it has many forms. The conscious and unconscious motivations associated with sexual attraction are diverse even among people of the same sex and orientation. Myriad experiences (and subjective interpretations of those experiences) could interact to lead different people to the same relative degree of sexual attraction to men or to women. Different people could be sexually attracted to men for different reasons; for example, there is no a priori reason that everyone attracted to men should share some particular brain structure.

Indeed, the notion that gay men are feminized and lesbians masculinized may tell us more about our culture than about the biology of erotic responsiveness. Some

Greek myths held that heterosexual rather than homosexual desire had intersex origins: those with predominately same-sex desires were considered the most manly of men and womanly of women. In contrast, those who desired the opposite sex supposedly mixed masculine and feminine in their being. Classical culture celebrated the homosexual exploits of archetypically masculine heroes such as Zeus, Hercules, and Julius Caesar. Until a decade ago (when missionaries repudiated the practice), boys among the Sambia of New Guinea would form attachments to men and fellate them; no one considered that behavior a female trait. Indeed, the Sambia believed ingesting semen to be necessary for attaining strength and virility.

But there is a more tangible problem for this intersex assumption: the traits of which homosexuals ostensibly have opposite-sex versions have not been conclusively shown to differ between men and women. Of the many supposed sex differences in the human brain reported over the past century, only one has proved consistently replicable: brain size varies with body size. Thus, men tend to have slightly larger brains than women. This situation contrasts sharply with that for other animals, where many researchers have consistently demonstrated a variety of sex differences.

If brains are indeed wired or otherwise programmed for sexual orientation, what forces are responsible? Three possibilities come into play: the direct model of biological causation asserts that genes, hormones, or other factors act directly on the developing brain, probably before birth, to wire it for sexual orientation. Alternatively, the social learning model suggests that biology provides a blank slate of neural circuitry on which experience inscribes orientation. In the indirect model, biological factors do not wire the brain for orientation; instead they predispose individuals toward certain personality traits that influence the relationships and experiences that ultimately shape sexuality.

During past decades, much of the speculation about biology and orientation focused on the role of hormones. Workers once thought an adult's androgen and estrogen levels determined orientation, but this hypothesis withered for lack of support. Researchers have since pursued the notion that hormones wire the brain for sexual orientation during the prenatal period.

According to this hypothesis, high prenatal androgen levels during the appropriate critical period cause heterosexuality in men and homosexuality in women. Conversely, low fetal androgen levels lead to homosexuality in men and heterosexuality in women. This hypothesis rests largely on the observation that in rodents early exposure to hormones determines the balance between male and female patterns of mating behaviors displayed by adults. Female rodents that were exposed to androgens early in development show more male-typical mounting behavior than do normal adult females. Males deprived of androgens by castration during the same critical period display a female mating posture called lordosis (bending of the back) when they are mounted.

Many researchers consider the castrated male rat that shows lordosis when mounted by another male to be homosexual (as is the female rat that mounts others). Lordosis, however, is little more than a reflex: the male will take the same posture when a handler strokes its back. Furthermore, the male that mounts another male is considered to be heterosexual, as is the female that displays lordosis when mounted by another female. Applying such logic to humans would imply that if two people of the same sex engaged in intercourse only one is homosexual—and which member of the couple it is depends on the positions they assume.

In addition to determining rodent mating patterns, early hormonal exposure determines whether an animal's brain can regulate normal ovarian function. A male rat's brain cannot respond to estrogen by triggering a chain of events, called positive feedback, that culminates in the abrupt increase of luteinizing hormone in the bloodstream, which in turn triggers ovulation. Some researchers reasoned from this fact to the idea that homosexual men (whose brains they allege to be insufficiently masculinized) might have a stronger positive-feedback reaction than do heterosexual men.

Two laboratories reported that this was the case, but carefully designed and executed studies, most notably those of Luis J. G. Gooren of the Free University in Amsterdam, disproved those findings. Furthermore, the feedback mechanism turns out to be irrelevant to human sexual orientation: workers have since found that the positive-feedback mechanism is not sexually dimorphic in primates, including humans. If this mechanism is indistinguishable in men and women, it is illogical to suggest that it should be "feminized" in gay men.

Moreover, a corollary of the expectation that luteinizing hormone responses should be feminized in homosexual men is that they should be "masculinized" in lesbians. If that were true, homosexual women would neither menstruate nor bear children. The overwhelming proportion of lesbians with normal menstrual cycles and the growing number of openly lesbian mothers attest to the fallacy of that idea.

If the prenatal hormonal hypothesis were correct, one might expect that a large proportion of men with medical conditions known to involve prenatal androgen deficiency would be homosexual, as would be women exposed prenatally to excess androgens. That is not the case.

Because androgens are necessary for development of normal external genitals in males, the sex of affected individuals may not be apparent at birth. Males may be born with female-appearing genitals, and females with male-appearing ones. These individuals often require plastic surgery to construct normal-appearing genitals, and the decision to raise them as boys or as girls is sometimes based not on genetic sex but on the possibilities for genital reconstruction.

Research into the sexual orientation of such individuals tends to support the social learning model. Regardless of their genetic sex or the nature of their prenatal hormonal exposure, they usually become heterosexual with respect to the sex their parents raise them as, provided the sex assignment is made unambiguously before the age of three.

Nevertheless, some studies report an increase in homosexual fantasies or behavior among women who were exposed to androgens as fetuses. In accordance with the notion of direct biological effects, these studies are often interpreted as evidence that prenatal androgen exposure wires the brain for sexual attraction to women. The neurobiologist and feminist scholar Ruth H. Bleier has offered an alternative interpretation. Rather than reflecting an effect of masculinizing hormones on the sexual differentiation of the brain, the adaptations of prenatally masculinized women may reflect the impact of having been born with masculinized genitalia or the knowledge that they had been exposed to aberrant levels of sex hormones during development. "Gender must seem a fragile and arbitrary construct," Bleier concluded, "if it depends upon plastic surgery."

Stephen Jay Gould of Harvard University has written of the way that the search for brain differences related to sex and other social categories was for the most part discredited during the past century by anatomists who deluded themselves into believing that their brain measurements justified the social prejudice of their day. The search for sex differences in the human brain was revitalized in the late 1970s, when Roger A. Gorski's team at the University of California at Los Angeles discovered a group of cells in the preoptic part of the rat hypothalamus that was much larger in males than in females. The researchers designated this cell group the sexually dimorphic nucleus of the preoptic area (SDN-POA). The preoptic area has long been implicated in the regulation of sexual behavior.

Like the sex differences in mating behaviors and luteinizing hormone regulatory mechanisms, the difference in the size of the SDN-POA was found to result from differences in early exposure to androgens. Shortly thereafter, Bleier and I, working at the University of Wisconsin at Madison, examined the hypothalamus of several rodent species and found that the SDN-POA is only one part of a sexual dimorphism involving several additional hypothalamic nuclei.

Three laboratories have recently sought sexually dimorphic nuclei in the human hypothalamus. Laura S. Allen, working in Gorski's lab, identified four possible candidates as potential homologues of the rat's SDN-POA and designated them as the interstitial nuclei of the anterior hypothalamus (INAH1–INAH4). Different labo-

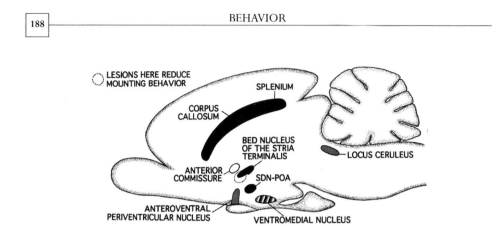

Sexually dimorphic nucleus of the preoptic area (SDN-POA) in the rat brain is among the regions whose size varies between males and females. Attempts to find an analogous cell group in humans have met with varying success [see table below]; some nuclei have not even been confirmed to exist in other rodents. Regions larger in males are shaded in black, and those larger in females are shaded in gray. (Patricia J. Wynne)

ratories that have measured these nuclei, however, have produced conflicting results: Dick F. Swaab's group at the Netherlands Institute for Brain Research in Amsterdam, for example, found INAH1 to be larger in men than in women, whereas Allen found no difference in that nucleus but reported that INAH2 and INAH3 were larger in men. Most recently, LeVay found no sex difference in either INAH1 or INAH2 but corroborated Allen's finding of a larger INAH3 in men. LeVay also reported that INAH3 in homosexual men tends to be small, like that of women. (Neurologist Clifford Saper of Harvard and I are in the process of measuring the interstitial nuclei; at present, we have no definitive results.)

RESEARCHERS	BRAIN REGION			
	INAH1	INAH2	INAH3	INAH4
Swaab and Fliers, 1985	Larger in men	Not studied	Not studied	Not studied
Allen et al., 1989	No sex difference	Larger in men than in some women	Larger in men	No sex difference
LeVay, 1991	No sex difference	No sex difference	Larger in heterosexual men than in women or homosexual men	No sex difference

Hypothalamic nuclei are reported to be sites of sexual differences in humans. Yet speculations about the possible contribution of those nuclei to sexual orientation are premature because no differences between men and women have been conclusively demonstrated in these regions.

LeVay's study has been widely interpreted as strong evidence that biological factors directly wire the brain for sexual orientation. Several considerations militate against that conclusion. First, his work has not been replicated, and human neuroanatomical studies of this kind have a very poor track record for reproducibility. Indeed, procedures similar to those LeVay used to identify the nuclei have previously led researchers astray.

Manfred Gahr, now at the Max Planck Institute for Animal Physiology in Seewiesen, Germany, used a cell-staining technique similar to LeVay's to observe what appeared to be seasonal variations in the size of a nucleus involved in singing in canaries. Two more specific staining methods, however, revealed that the size of the nucleus did not change. Gahr suggested that the less specific method might have been influenced by seasonal hormonal variations that altered the properties of the cells in the nucleus.

Furthermore, in LeVay's published study, all the brains of gay men came from AIDS patients. His inclusion of a few brains from heterosexual men with AIDS did not adequately address the fact that at the time of death virtually all men with AIDS have decreased testosterone levels as the result of the disease itself or the side effects of particular treatments. To date, LeVay has examined the brain of only one gay man who did not die of AIDS. Thus, it is possible that the effects on the size of INAH3 that he attributed to sexual orientation were actually caused by the hormonal abnormalities associated with AIDS. Work by Deborah Commins and Pauline I. Yahr of the University of California at Irvine supports precisely this hypothesis. The two found that the size of a structure in mongolian gerbils apparently comparable to the SDN-POA varies with the amount of testosterone in the bloodstream.

A final problem with the popular interpretation of LeVay's study is that it is founded on an imprecise analysis of the relevant animal research. LeVay has suggested that INAH3, like the rat's SDN-POA, is situated in a region of the hypothalamus known to participate in the generation of male sexual behavior. Yet studies in a variety of species have consistently shown that the precise hypothalamic region involved in male sexual behavior is not the one occupied by these nuclei. Indeed, Gorski and Gary W. Arendash, now at the University of South Florida, found that destroying the SDN-POA on both sides of a male rat's brain did not impair sexual behavior.

Jefferson C. Slimp performed experiments in Robert W. Goy's laboratory at the Wisconsin Regional Primate Research Center (shortly before I joined that group) that suggested that the precise region involved in sexual behavior in male rhesus monkeys is located above the area comparable to that occupied by INAH3 in humans. Males with lesions in that region mounted females less frequently than they did before being operated on, but their frequency of masturbation did not change. Although some have taken these observations to mean that the lesions selectively decreased heterosexual drive, their conclusion is unwarranted; male monkeys pressed a lever for access to females more often after their operations than before. Unfortunately, these males had no opportunity to interact with other males, and so the study tells us nothing about effects on homosexual as opposed to heterosexual motivation or behavior.

Interstitial hypothalamic nuclei are not the only parts of the brain to have come under scrutiny for links to sexual orientation. Neuroanatomists have also reported potentially interesting differences in regions not directly involved in sexual behaviors. Swaab and his co-worker Michel A. Hofman found that another hypothalamic nucleus, the suprachiasmatic nucleus, is larger in homosexual than in heterosexual men. The size of this structure, however, does not vary with sex, and so even if this finding can be replicated it would not support the assumption that homosexuals have intersexed brains.

Allen of U.C.L.A., meanwhile, has reported that the anterior commissure, a structure that participates in relaying information from one side of the brain to the other, is larger in women than in men. More recently, she concluded that the anterior commissure of gay men is feminized—that is, larger than in heterosexual men. Steven Demeter, Robert W. Doty, and James L. Ringo of the University of Rochester, however, found just the opposite: anterior commissures larger in men than in women. Furthermore, even if Allen's findings are correct, the size of the anterior commissure alone would say nothing about an individual's sexual orientation. Although she found a statistically significant difference in the average size of the commissure of gay men and heterosexual men, 27 of the 30 homosexual men in her study had anterior commissures within the same size range as the 30 heterosexual men with whom she compared them.

Some researchers have turned to genetics instead of brain structure in the search for a biological link to sexual orientation. Several recent studies suggest that the brothers of homosexual men are more likely to be homosexual than are men without gay brothers. Of these, only the study by J. Michael Bailey of Northwestern University and Richard C. Pillard of Boston University included both nontwin biological brothers and adopted (unrelated) brothers in addition to identical and fraternal twins.

Their investigation yielded paradoxical results: some statistics support a genetic hypothesis, and others refute it. Identical twins were most likely to both be gay; 52 percent were concordant for homosexuality, as compared with 22 percent of fraternal twins. This result would support a genetic interpretation because identical twins share all of their genes, whereas fraternal twins share only half of theirs. Nontwin brothers of homosexuals, however, share the same proportion of genes as fraternal twins; however, only 9 percent of them were concordant for homosexuality. The genetic hypothesis predicts that their rates should be equal.

Moreover, Bailey and Pillard found that the incidence of homosexuality in the adopted brothers of homosexuals (11 percent) was much higher than recent estimates for the rate of homosexuality in the population (1 to 5 percent). In fact it was equal to the rate for non-twin biological brothers. This study clearly challenges a simple genetic hypothesis and strongly suggests that environment contributes significantly to sexual orientation.

Two of three other recent studies also detected an increased rate of homosexuality among the identical as opposed to fraternal twins of homosexuals. In every case, however, the twins were reared together. Without knowing what developmental experiences contribute to sexual orientation—and whether those experiences are more similar between identical twins than between fraternal twins—the effects of common genes and common environments are difficult to disentangle. Resolving this issue requires studies of twins raised apart.

Indeed, perhaps the major finding of these heritability studies is that despite having all of their genes in common and having prenatal and postnatal environments as close to identical as possible, approximately half of the identical twins were nonetheless discordant for orientation. This finding underscores just how little is known about the origins of sexual orientation.

Dean H. Hamer's team at the National Institutes of Health has found the most direct evidence that sexual orientation may be influenced by specific genes. The team focused on a small part of the X chromosome known as the Xq28 region, which contains hundreds of genes. Women have two X chromosomes and so two Xq28 regions, but they pass a copy of only one to a son (who has a single X chromosome). The theoretical probability of two sons receiving a copy of the same Xq28 from their mother is thus 50 percent. Hamer found that of his 40 pairs of gay siblings, 33 instead of the expected 20 had received the same Xq28 region from their mother.

Hamer's finding is often misinterpreted as showing that all 66 men from these 33 pairs shared the same Xq28 sequence. That is quite different from what the study showed: each member of the 33 concordant pairs shared his Xq28 region only with his brother—not with any of the other 32 pairs. No single, specific Xq28 sequence (a putative "gay gene") was identified in all 66 men.

Unfortunately, Hamer's team did not examine the Xq28 region of its gay subjects' heterosexual brothers to see how many shared the same sequence. Hamer suggests that inclusion of heterosexual siblings would have confounded his analysis because the gene associated with homosexuality might be "incompletely penetrant"—that is to say, heterosexual men could carry the gene without expressing it.

In other words, inclusion of heterosexual brothers might have revealed that something other than genes is responsible for sexual orientation.

Finally, Neil J. Risch of Yale University, one of the developers of the statistical techniques that Hamer used, has questioned whether Hamer's results are statistically significant. Risch has argued that until we have more details about the familial clustering of homosexuality, the implications of studies such as Hamer's will remain unclear.

Studies that mark homosexuality as a heritable trait (assuming that they can be replicated) do not say anything about how that heritability might operate. Genes in themselves specify proteins, not behavior, or psychological phenomena. Although we know virtually nothing about how complex psychological phenomena are embodied in the brain, it is conceivable that particular DNA sequences might somehow cause the brain to be wired specifically for homosexual orientation. Significantly, however, heritability requires no such mechanism.

Instead particular genes might influence personality traits that could in turn influence the relationships and subjective experiences that contribute to the social learning of sexual orientation. One can imagine many ways in which a temperamental difference could give rise to different orientations in different environments.

The *Achillea* plant serves as a useful metaphor: genetic variations yield disparate phenotypes depending on elevation. The altitude at which a cutting of *Achillea* grows does not have a linear effect on the plant's growth, however, nor is the plant limited to a single attribute. Height, number of leaves and stems, and branching pattern are all affected [*see illustration on page 193*]. If a plant can display such a complex response to its environment, then what of a far more complex organism that can modify its surroundings at will?

The possible interaction between genes and environment in the development of sexual orientation can be sketched here only in the most oversimplified of ways. For example, many researchers believe aversion to rough-and-tumble play in boys is moderately predictive of homosexual development. (Direct-model theorists argue this aversion is merely the childhood expression of a brain that has been wired for homosexuality.) Meanwhile psychoanalysts have noted that of those gay men who seek therapy, many report having had poor rapport with their fathers. They thus suggest that an impaired father-son relationship leads to homosexuality.

One could combine these observations to speculate that a genetically based aversion to rough-and-tumble play in boys could impair rapport with fathers who demand that they adhere to rigid sex-role stereotypes. Fathers who made no such demands would maintain a rapport with their sons. As a result, the hypothetical gene in question could affect sexual orientation in some cases but not in others. Even

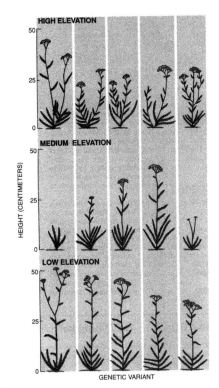

Cuttings from Achillea *plants have the same genes, yet they develop in significantly different ways depending on their environment. Furthermore, knowing how the five genetic variants above differ in one environment does not help predict their traits in another one. That plants can display such a complex response to their surroundings makes clear the illogic of expecting direct, easily predictable links between human genes and as diffuse a trait as sexual orientation. (Patricia J. Wynne)*

such a reductionist example (based on traits that reflect cultural stereotypes rather than biology) shows how neither temperament nor family environment might be decisive. Studies focusing on either one or the other would yield inconclusive results.

These speculations reemphasize how far researchers must go before they understand the factors—both biological and experiential—that contribute to sexual orientation. Even if the size of certain brain structures does turn out to be correlated with sexual orientation, current understanding of the brain is inadequate to explain how such quantitative differences could generate qualitative differences in a psychological phenomenon as complex as sexual orientation. Similarly, confirmation of genetic research purporting to show that homosexuality is heritable make clear neither what is inherited nor how it influences sexual orientation. For the foreseeable future, then, interpretation of these results will continue to hinge on assumptions of questionable validity.

While attempts to replicate these preliminary findings continue, researchers and the public must resist the temptation to consider them in any but the most tentative fashion. Perhaps more important, we should also be asking ourselves why we as a society are so emotionally invested in this research. Will it—or should it—make any difference in the way we perceive ourselves and others or how we live our lives and allow others to live theirs? Perhaps the answers to the most salient questions in this debate lie not within the biology of human brains but rather in the cultures those brains have created.

—May 1994

THE NEUROBIOLOGY
OF FEAR

*Researchers are teasing apart the neurochemical processes
that give rise to different fears in monkeys. The results may
lead to new ways to treat anxiety in humans*

Ned H. Kalin

Over the years, most people acquire a repertoire of skills for coping with a range of frightening situations. They will attempt to placate a vexed teacher or boss and will shout and run when chased by a mugger. Some individuals, though, become over-whelmed in circumstances others would consider only minimally stressful: fear of ridicule might cause them to shake uncontrollably when called on to speak in a group, or terror of strangers might lead them to hide at home, unable to work or shop for groceries. Why do certain people fall prey to excessive fear?

At the University of Wisconsin at Madison, my colleague Steven E. Shelton and I are addressing this problem by identifying specific brain processes that regulate fear and its associated behaviors. Despite the availability of noninvasive imaging techniques, such information is still extremely difficult to obtain in humans. Hence, we have turned our attention to another primate, the rhesus monkey (*Macaca mulatta*). These animals undergo many of the same physiological and psychological developmental stages that humans do, but in a more compressed time span. As we gain more insight into the nature and operation of neural circuits that modulate fear in monkeys, it should be possible to pinpoint the brain processes that cause inordinate anxiety in people and to devise new therapies to counteract it.

Effective interventions would be particularly beneficial if they were applied at an early age. Growing evidence suggests overly fearful youngsters are at high risk for later emotional distress. Jerome Kagan and his colleagues at Harvard University have shown, for example, that a child who is profoundly shy at the age of two years is more likely than a less inhibited child to suffer from anxiety and depression later in life.

This is not to say these ailments are inevitable. But it is easy to see how excessive fear could contribute to a lifetime of emotional struggle. Consider a child who is deeply afraid of other children and is therefore taunted by them at school. That youngster might begin to feel unlikable and, in turn, to withdraw further. With time the growing child could become mired in a vicious circle leading to isolation, low self-esteem, underachievement, and the anxiety and depression noted by Kagan.

There are indications that unusually fearful children might also be prone to physical illness. Many youngsters who become severely inhibited in unfamiliar situations chronically overproduce stress hormones, including the adrenal product cortisol. In times of threat, these hormones are critical. They ensure that muscles have the energy needed for "fight or flight." But some evidence indicates long-term elevations of stress hormones may contribute to gastric ulcers and cardiovascular disease.

Further, through unknown mechanisms, fearful children and their families are more likely than others to suffer from allergic disorders. Finally, in rodents and nonhuman primates, persistent elevation of cortisol has been shown to increase the vulnerability of neurons in the hippocampus to damage by other substances; this brain region is involved in memory, motivation, and emotion. Human neurons probably are affected in a similar manner, although direct evidence is awaited.

When we began our studies about 10 years ago, Shelton and I knew we would first have to find cues that elicit fear and identify behaviors that reflect different types of anxiety. With such information in hand, we could proceed to determine the age at which monkeys begin to match defensive behaviors selectively to specific cues. By also determining the parts of the brain that reach maturity during the same time span, we could gain clues to the regions that underlie the regulation of fear and fear-related behavior.

The experiments were carried out at the Wisconsin Regional Primate Research Center and the Harlow Primate Laboratory, both at the University of Wisconsin. We discerned varied behaviors by exposing monkeys between six and 12 months old to three related situations. In the alone condition, an animal was separated from its mother and left by itself in a cage for 10 minutes. In the no-eye-contact condition, a person stood motionless outside the cage and avoided looking at the solitary infant. In the stare condition, a person was again present and motionless but, assuming a neutral expression, peered directly at the animal. These conditions are no more frightening than those primates encounter frequently in the wild or those human infants encounter every time they are left at a day-care center.

■ THREE TYPICAL FEAR BEHAVIORS

In the alone condition, most monkeys became very active and emitted frequent "ooo" calls. These fairly melodious sounds are made with pursed lips. They start at a low pitch, become higher, and then fall. More than 30 years ago Harry F. Harlow, then at Wisconsin, deduced that when an infant monkey is separated from its mother, its primary goal is affiliative—that is, it yearns to regain the closeness and sense of security provided by nearness to the parent. Moving about and cooing help to draw the mother's attention.

In contrast, in the more frightening no-eye-contact situation, the monkeys reduced their activity greatly and sometimes "froze," remaining completely still for prolonged periods. When an infant spots a possible predator, its goal shifts from attracting the mother to becoming inconspicuous. Inhibiting motion and freezing—common responses in many species—reduce the likelihood of attack.

If the infant perceives that it has been detected, its aim shifts again, to warding off an attack. And so the stare condition evoked a third set of responses. The monkeys made several hostile gestures, among them "barking" (forcing air from the abdomen through the vocal cords to emit a growllike sound), staring back, producing so-called threat faces, baring their teeth, and shaking the cage. Sometimes the animals mixed the threatening displays with submissive ones, such as fear grimaces, which look something like wary grins, or grinding of the teeth. In this condition, too, cooing increased over the amount heard when the animals were alone. (As will be seen, we have recently come to think the cooing displayed in the stare condition may serve a somewhat different function than it does in the alone situation.)

Monkeys, by the way, are not unique in becoming aroused by stares and in using them reciprocally to intimidate predators. Animals as diverse as crabs, lizards, and birds all perceive staring as a threat. Some fishes and insects have evolved protec-

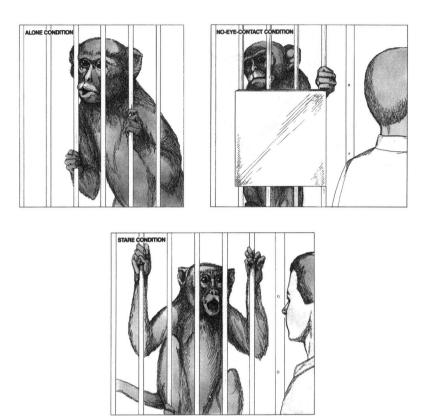

Three experimental conditions elicit distinct fear-related behaviors in rhesus monkeys older than about two months. When isolated in a cage (top left), youngsters become quite active and emit "coo" sounds to attract their mothers. If a human appears but avoids eye contact (top right), the monkeys try to evade the discovery, such as by staying completely still (freezing) or hiding behind their food bin. If the intruder stares at the animals (bottom), they become aggressive. (Carol Donner)

tive spots that resemble eyes; these spots either avert attacks completely or redirect them to nonvital parts of the body. In India, field-workers wear face masks behind their heads to discourage tigers from pouncing at their backs. Studies of humans show that we, too, are sensitive to direct gazes: brain activity increases when we are stared at, and people who are anxious or depressed tend to avoid direct eye contact.

Having identified three constellations of defensive behaviors, we set about determining when infant monkeys first begin to apply them effectively. Several lines of work led us to surmise that the ability to make such choices emerges sometime around an infant's two-month birthday. For instance, rhesus mothers generally permit children to venture off with their peers at that time, presumably because the adults are now confident that the infants can protect themselves reasonably well. We also knew that by about 10 weeks of age infant monkeys respond with different emotions to specific expressions on the faces of other monkeys—a sign that at least some of the innate wiring or learned skills needed to discriminate threatening cues are in place.

To establish the critical period of development, we examined four groups of monkeys ranging in age from a few days to 12 weeks old. We separated the babies from their mothers and let them acclimate to an unfamiliar cage. Then we exposed them to the alone, no-eye-contact, and stare conditions. All sessions were video-taped for analysis.

We found that infants in the youngest group (newborns to two-week-olds) engaged in defensive behaviors. But they lacked some motor coordination and seemed to act randomly, as if they were oblivious to the presence or gaze of the human intruder. Babies in our two intermediate-age groups had good motor control, but their actions seemed unrelated to the test condition. This finding meant motor control was not the prime determinant of selective responding.

Only animals in our oldest group (nine- to 12-week-olds) conducted themselves differently in each situation, and their reactions were both appropriate and identical to those of mature monkeys. Nine to 12 weeks, then, is the critical age of the appearance of a monkey's ability to adaptively modulate its defensive activity to meet changing demands.

Studies by other workers, who primarily examined rodents, suggested that three interconnected parts of the brain regulate fearfulness. We suspected that these regions become functionally mature during the nine- to 12-week period and thus give rise to the selective reactivity we observed. One of these regions is the prefrontal cortex, which takes up much of the outer and side areas of the cerebral cortex in the frontal lobe [see illustration on page 200]. A cognitive and emotional area, the prefrontal cortex is thought to participate in the interpretation of sensory stimuli and is probably a site where the potential for danger is assessed.

The second region is the amygdala, a part of a primitive area in the brain called the limbic system (which includes the hippocampus). The limbic system in general and the amygdala in particular have been implicated in generating fear.

The final region is the hypothalamus. Located at the base of the brain, it is a constituent of what is called the hypothalamic-pituitary-adrenal system. In response to stress signals from elsewhere in the brain, such as the limbic system and other cortical regions, the hypothalamus secretes corticotropin-releasing hormone. This small protein spurs the pituitary gland, located just below the brain, to secrete adrenocorticotropic hormone (ACTH), which prods the adrenal gland to release cortisol, which prepares the body to defend itself.

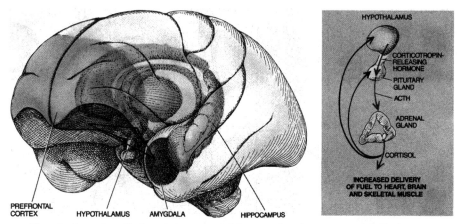

Three brain regions that are interconnected by neural pathways are critically impor-
tant in regulating fear-related behaviors. The prefrontal cortex participates in assess-
ing danger. The amygdala is a major constituent of the emotion-producing limbic sys-
tem. And the hypothalamus, in response to signals from the prefrontal cortex,
amygdala, and hippocampus, directs the release of hormones that support motor re-
sponses to perceived threats. (Carol Donner)

In neuroanatomic data collected in other laboratories, we found support for our suspicion that maturation of these brain regions underlies selective responding in the nine- to 12-week period. For instance, during this time the formation of synapses (contact points between neurons) has been shown to reach its peak in the prefrontal cortex and the limbic system (including the amygdala), as well as in the motor and visual cortices and other sensory areas. Patricia S. Goldman-Rakic of Yale University has also established that as the prefrontal cortex matures in rhesus monkeys, the ability to guide behavior based on experience emerges. This skill is necessary if one is to contend successfully with danger.

Maturation of the prefrontal cortex likewise seems important for enabling humans to distinguish among threatening cues. Harry T. Chugani and his co-workers at the University of California at Los Angeles have shown that activity in the prefrontal cortex increases when human offspring are seven to 12 months of age. During this span—which appears to be analogous to the time when monkeys begin to respond selectively to fear—children begin to display marked fear of strangers. They also become adept at what is called social referencing; they regulate their level of fear based on interpreting the expressions they observe on a parent's face.

But what of the hypothalamus, the third brain region we assumed could participate in regulating fear-related behavior? Published research did not tell us much about its development or about the development of the complete hypothalamic-pituitary-adrenal system in monkeys. Our own investigations, however, revealed that the full system matures in parallel with that of the prefrontal cortex and the limbic system.

In these studies, we used the pituitary hormone ACTH as a marker of the system's function. We again examined four groups of infants aged a few days to 12 weeks. From each subject, we measured ACTH levels in blood drawn while the youngster was with its mother. This reading provided a baseline. We also measured ACTH levels in blood samples obtained 20 minutes after the infant was separated from its parent. Hormonal levels rose in all four age groups during separation, but they jumped profoundly only in the oldest (nine- to 12-week-old) moneys.

The relatively weak response in the younger animals, particularly in those under two weeks old, is consistent with findings in rat pups, whose stress hormone response is also blunted during the first two weeks of life. The development of the rodent and primate stress hormone system may well be delayed during early life to protect young neurons from the potentially damaging effects of cortisol.

Assured that the hypothalamic-pituitary-adrenal system becomes functionally mature by nine to 12 weeks, we pressed the inquiry forward to determine whether levels of cortisol and ACTH might partly account for individual differences in defensive behavior. We were also curious to know whether the responses of the infants resembled those of their mothers; a correspondence would indicate that further analyses of mothers and their infants could help reveal the relative contributions of inheritance and learning to fearfulness. We mainly examined the propensity for freezing, which we had earlier found was a stable trait in our subjects.

■ MATURING OF THE FEAR RESPONSE

In one set of studies, we measured baseline levels of cortisol in monkeys four months to a year old and then observed how much time the youngsters froze in the no-eye-contact condition. Monkeys that started off with relatively low levels of cortisol froze for shorter periods than did their counterparts with higher cortisol levels—a pattern we also noted in separate studies of adult females. In other studies, we observed that as youngsters pass through their first year of life, they become progressively like their mothers hormonally and behaviorally. By the time infants are about five months old, their stress-induced rises in ACTH levels parallel those of the mother. And by the time they are a year old, the duration of freezing in the no-eye-contact condition also corresponds to that of the mother.

Strikingly, some of these results echoed those obtained in humans. Extremely inhibited children often have parents who suffer from anxiety. Moreover, Kagan and his colleagues have found that basal cortisol levels are predictive of such children's reaction to a frightening situation. They measured cortisol concentration in saliva of youngsters at home (where they are presumably most relaxed) and then observed the children confronting an unfamiliar situation in the laboratory; high basal cortisol levels were associated with greater inhibition in the strange setting.

These similarities between humans and monkeys again imply that monkeys are reasonable models of human emotional reactivity. The link between basal cortisol levels and duration of freezing or inhibition suggests as well that levels of stress hormones influence how appropriately animals and people behave in the face of fear. (This effect may partly be mediated by the hippocampus, where the concentration of cortisol receptors is high.) And the likeness of hormonal and behavioral responses in mothers and infants implies that genetic inheritance might predispose some individuals to extreme fearfulness, although we cannot rule out the contribution of experience.

No one can yet say to what extent the activity of the hypothalamic-pituitary-adrenal system controls, and is controlled by, other brain regions that regulate the choice of defensive behavior. We have, however, begun to identify distinct neurochemical circuits, or systems, in the brain that affect different behaviors. The two systems we have studied most intensely seemed at first to have quite separate functions. But more recent work implies that the controls on defensive behavior are rather more complicated than the original analyses implied.

We gathered our initial data by treating six- to 12-month-old monkeys with two different classes of neuroactive chemicals—opiates (morphinelike substances) and benzodiazepines (chemicals that include the antianxiety drug diazepam, or Valium). We chose to look at opiates and benzodiazepines because neurons that release or take up those chemicals are abundant in the prefrontal cortex, the amygdala, and the hypothalamus. The opiates are known to have natural, or endogenous, counterparts, called endorphins and enkephalins, that serve as neurotransmitters; after the endogenous chemicals are released by certain neurons, they bind to receptor molecules on other nerve cells and thereby increase or decrease nerve cell activity. Receptors for benzodiazepines have been identified, but investigators are still trying to isolate endogenous benzodiazepinelike molecules.

■ SELECTIVE DRUG EFFECTS

Once again, our subjects were exposed to the alone, no-eye-contact, and stare conditions. We delivered the drugs before the infants were separated from their mothers and then recorded the animals' behavior. Morphine decreased the amount of cooing normally displayed in the alone and stare conditions. Conversely, cooing

	COOING	FREEZING	BARKING
MORPHINE (OPIATE)	DECREASES	NO EFFECT	NO EFFECT
NALOXONE (OPIATE BLOCKER)	INCREASES	NO EFFECT	NO EFFECT
DIAZEPAM (BENZODIAZEPINE)	NO EFFECT	DECREASES	DECREASES

Effects on cooing, freezing, and barking were evaluated some years ago for three drugs that act on neurons responsive to opiates (top two rows) or to benzodiazepines (bottom row). The results implied that opiate-sensitive pathways in the brain control affiliative behaviors (those that restore closeness to the mother, as cooing often does), whereas benzodiazepine-sensitive pathways control responses to immediate threats (such as freezing and barking). Newer evidence generally supports this conclusion but adds some complexity to the picture. (Carol Donner)

was increased by naloxone, a compound that binds to opiate receptors but blocks the activity of morphine and endogenous opiates. Yet morphine and naloxone had no influence on the frequency of stare-induced barking and other hostile behaviors, nor did they influence duration of freezing in the no-eye-contact situation. We concluded that opiate-using neural pathways primarily regulate affiliative behaviors (such as those induced by distress over separation from the mother), but those pathways seem to have little power over responses to direct threats.

The benzodiazepine we studied—diazepam—produced a contrary picture. The drug had no impact on cooing, but it markedly reduced freezing, barking, and other hostile gestures. Thus, benzodiazepine-using pathways seemed primarily to influence responses to direct threats but to have little power over affiliative behavior.

We still think the opiate and benzodiazepine pathways basically serve these separate functions. Nevertheless, the simple model we initially envisioned grew more interesting as we investigated two additional drugs: a benzodiazepine called alprazolam (Xanax) and a compound called beta-carboline, which binds to benzodiazepine receptors but elevates anxiety and typically produces effects opposite to those of diazepam and its relatives.

When we administered alprazolam in doses that lower anxiety enough to decrease freezing, this substance, like diazepam, minimalized hostility in the threatening, stare condition. And beta-carboline enhanced hostility. No surprises here. Yet, unlike diazepam, these drugs modulated cooing, which we had considered to be an affiliative (opiate-controlled), not a threat-related (benzodiazepine-controlled), behavior. Moreover, both these compounds decreased cooing. We cannot explain the similarity of effect, but we have some ideas about why drugs that act on benzodiazepine receptors might influence cooing.

It may be that, contrary to our early view, benzodiazepine pathways can in fact regulate affiliative behavior. We favor a second interpretation, however. Cooing displayed in the stare condition may not solely reflect an affiliative need (a desire for mother's comfort); at times, it may also be an urgent, threat-induced plea for immediate help. One behavior, then, might serve two different functions and be controlled by different neurochemical pathways. (This conclusion was strengthened for me recently, when I tried to photograph a rhesus infant that had become separated from its mother in the wild—where we are now initiating additional studies. Its persistent, intense coos attracted the mother, along with a pack of protectors. The strategy worked: I retreated rapidly.)

More generally, our chemical studies lead us to suspect that the opiate- and benzodiazepine-sensitive circuits both operate during stress; the relative degree of activity changes with the characteristics of a worrisome situation. As the contribution of each pathway is altered, so, too, are the behaviors that appear.

Exactly how neurons in the opiate and benzodiazepine pathways function and how they might cooperate are unclear. But one plausible scenario goes like this: When a young monkey is separated from its mother, opiate-releasing and, consequently, opiate-sensitive neurons become inhibited. Such inhibition gives rise to yearning for the mother and a generalized sense of vulnerability. This reduction of activity in opiate-sensitive pathways enables motor systems in the brain to produce cooing. When a potential predator appears, neurons that secrete endogenous benzodiazepines become suppressed to some degree. This change, in turn, leads to elevated anxiety and the appearance of behaviors and hormonal responses that accompany fear. As the sense of alarm grows, motor areas prepare for fight or flight. The benzodiazepine system may also influence the opiate system, thereby altering cooing during threatening situations.

We are now refining our model of brain function by testing other compounds that bind to opiate and benzodiazepine receptors. We are also examining behavioral responses to substances, such as the neurotransmitter serotonin, that act on other receptors. (Serotonin receptors occur in many brain regions that participate in the expression of fear.) And we are studying the activities of substances that directly control stress hormone production, including corticotropin-releasing hormone, which is found throughout the brain, not solely in the hypothalamus.

In collaboration with Richard J. Davidson, here at Wisconsin, Shelton and I have recently identified at least one brain region where the benzodiazepine system exerts its effects. Davidson had shown that the prefrontal cortex of the right hemisphere is usually active in extremely inhibited children. We therefore wondered whether we would see the same asymmetry in frightened monkeys and whether drugs that reduced fear-related behavior in the animals would dampen right frontal activity.

This time we used mild restraint as a stress. As we anticipated, neuronal firing rose more in the right frontal cortex than in the left. Moreover, when we delivered diazepam in doses we knew lowered hostility, the drug returned the restraint-induced electrical activity to normal. In other words, the benzodiazepine system influences defensive behavior at least in part by acting in the right prefrontal cortex.

These findings have therapeutic implications. If human and monkey brains do operate similarly, our data would suggest that benzodiazepines might be most helpful in those adults and children who exhibit elevated electrical activity in the right prefrontal cortex. Because of the potential for side effects, many clinicians are cautions about delivering antianxiety medications to children over a long time. But administration of such drugs during critical periods of brain development might prove sufficient to alter the course of later development. It is also conceivable that behavioral training could teach extremely inhibited youngsters to regulate benzodiazepine-sensitive systems without having to be medicated. Alternatively, by screening compounds that are helpful in monkeys, investigators might discover new drugs that are quite safe for children. As the workings of other fear-modulating neurochemical systems in the brain are elucidated, similar strategies could be applied to manage those circuits.

Our discovery of cues that elicit three distinct sets of fear-related behaviors in rhesus monkeys has thus enabled us to gain insight into the development and regulation of defensive strategies in these animals. We propose that the opiate and benzodiazepine pathways in the prefrontal cortex, the amygdala, and the hypothalamus play a major part in determining which strategies are chosen. And we are currently attempting to learn more about the way in which these and other neural circuits coordinate with one another. We have therefore laid the groundwork for deciphering the relative contributions of various brain systems to inordinate fear in humans. We can envision a time when treatments will be tailored to normalizing the specific signaling pathways that are disrupted in a particular child, thereby sparing that youngster enormous unhappiness later in life.

—May 1993

SEEKING THE
CRIMINAL ELEMENT

Scientists are homing in on social and biological risk factors
that they believe predispose individuals to criminal behavior.
The knowledge could be ripe with promise—or rife with danger

W. Wayt Gibbs

"Imagine you are the father of an eight-year-old boy," says psychologist Adrian Raine, explaining where he believes his 17 years of research on the biological basis of crime is leading. "The ethical dilemma is this: I could say to you, 'Well, we have taken a wide variety of measurements, and we can predict with 80 percent accuracy that your son is going to become seriously violent within 20 years. We can offer you a series of biological, social, and cognitive intervention programs that will greatly reduce the chance of his becoming a violent offender.' What do you do? Do you place your boy in those programs and risk stigmatizing him as a violent criminal even though there is a real possibility that he is innocent? Or do you say no to the treatment and run an 80 percent chance that your child will grow up to (a) destroy his life, (b) destroy your life, (c) destroy the lives of his brothers and sisters, and, most important, (d) destroy the lives of the innocent victims who suffer at his hands?"

For now, such a choice is purely hypothetical. Scientists cannot yet predict which children will become dangerously aggressive with anything like 80 percent accuracy. But increasingly, those who study the causes of criminal and violent behavior are looking beyond broad demographic characteristics such as age, race, and income level to factors in individuals' personality, history, environment, and physiology that seem to put them—and society—at risk. As sociologists reap the benefits of rigorous long-term studies and neuroscientists tug at the tangled web of relations between behavior and brain chemistry, many are optimistic that science will identify markers of maleficence. "This research might not pay off for 10 years, but in 10 years it might revolutionize our criminal justice system," asserts Roger D. Masters, a political scientist at Dartmouth College.

■ PREVENTIVE INTERVENTION

"With the expected advances, we're going to be able to diagnose many people who are biologically brain-prone to violence," claims Stuart C. Yudofsky, chair of the psychiatry department at Baylor College of Medicine and editor of the *Journal of Neuropsychiatry and Clinical Neurosciences.* "I'm not worried about the downside as much as I am encouraged by the opportunity to prevent tragedies—to screen people who might have high risk and to prevent them from harming someone else." Raine, Yudofsky, and others argue that in order to control violence, Americans should trade their traditional concept of justice based on guilt and punishment for a "medical model" based on prevention, diagnosis, and treatment.

But many scientists and observers do worry about a downside. They are concerned that some researchers underplay the enormous complexity of individual behavior and overstate scientists' ability to understand and predict it. They also fear that a society desperate to reduce crime might find the temptation to make premature use of such knowledge irresistible.

Indeed, the history of science's assault on crime is blemished by instances in which incorrect conclusions were used to justify cruel and unusual punishments. In the early 1930s, when the homicide rate was even higher than it is today, eugenics was in fashion. "The eugenics movement was based on the idea that certain mental illness and criminal traits were all inherited," says Ronald L. Akers, director of the Center for Studies in Criminology and Law at the University of Florida. "It was based on bad science, but they thought it was good science at the time." By 1931, 27 states had passed laws allowing compulsory sterilization of "the feeble-minded," the insane, and the habitually criminal.

Studies in the late 1960s—when crime was again high and rising—revealed that many violent criminals had an extra Y chromosome and thus an extra set of "male" genes. "It was a dark day for science in Boston when they started screening babies for it," recalls Xandra O. Breakefield, a geneticist at Massachusetts General Hospital. Subsequent studies revealed that although XYY men tend to score lower on IQ tests, they are not unusually aggressive.

Social science studies on the causes of crime have been less controversial, in part because they have focused more on populations than on individuals. But as consensus builds among criminologists on a few key facts, researchers are assembling these into prediction models that try to identify the juveniles most likely to lapse into delinquency and then into violent crime.

Perhaps their most consistent finding is that a very small number of criminals are responsible for most of the violence. One study, for example, tracked for 27 years 10,000 males born in Philadelphia in 1945; it found that just 6 percent of them committed 71 percent of the homicides, 73 percent of the rapes, and 69 percent of the aggravated assaults attributed to the group.

Preventing just a small fraction of adolescent males from degenerating into chronic violent criminals could thus make a sizable impact on the violent crime rate, which has remained persistently high since 1973 despite a substantial decline in property crime. (Females accounted for only 12.5 percent of violent crime in 1992.) "For every 1 percent that we reduce violence, we save the country $1.2 billion," Raine asserts.

The problem says Terrie E. Moffitt, a psychologist at the University of Wisconsin who is conducting long-term delinquency prediction studies, is that "a lot of adolescents participate in antisocial behavior"—87 percent, according to a survey of U.S. teens. "The vast majority desist by age 21," she says. The dangerous few "are buried within that population of males trying out delinquency. How do you pick them out? Our hypothesis is that those who start earliest are at highest risk."

Marion S. Forgatch of the Oregon Social Learning Center tested that hypothesis on 319 boys from high-crime neighborhoods in Eugene. At the November 1994 American Society of Criminology meeting, she reported her findings: boys who had been arrested by age 14 were 17.9 times more likely to become chronic offenders than those who had not, and chronic offenders were 14.3 times more likely to commit violent offenses. "This is a good way of predicting," she says.

■ FALSE POSITIVE ID

Good is a relative term. For if one were to predict that every boy in her study who was arrested early would go on to commit violent crimes, one would be wrong more than 65 percent of the time. To statisticians, those so misidentified are known as false positives. "All of these predictors have a lot of false positives—about 50 percent on average," says Akers, who recently completed a survey of delinquency prediction models. Their total accuracy is even lower, because the models also fail to identify some future criminals.

The risk factors that Akers says researchers have found to be most closely associated with delinquency are hardly surprising. Drug use tops the list, followed by family dysfunction, childhood behavior problems, deviant peers, poor school performance, inconsistent parental supervision and discipline, separation from parents, and poverty. Numerous other controlled studies have found that alcoholism, childhood abuse, low verbal IQ, and witnessing violent acts are also significant risk factors. Compared with violent behavior, however, all these experiences are exceedingly common. The disparity makes it very difficult to determine which factors are causes and which merely correlates.

The difference is important, notes Mark W. Lipsey of Vanderbilt University, because "changing a risk factor if it is not causal may have no impact," and the ultimate goal of prediction is to stop violence by intervening before it begins. Unfortunately, improvements in predictive models do not necessarily translate into effective intervention strategies. Lipsey recently analyzed how well some 500 delinquency treatment programs reduced recidivism. "The conventional wisdom that nothing works is just wrong," he concludes. But he concedes that "the net effect is modest"—on average, 45 percent of program participants were rearrested, versus 50 percent of those left to their own devices. Half of that small apparent improvement, he adds, may be the result of inconsistency in the methods used to evaluate the programs.

Some strategies do work better than others, Lipsey discovered. Behavioral programs that concentrated on teaching job skills and rewarding prosocial attitudes cut rearrest rates to about 35 percent. "Scared straight" and boot camp programs, on the other hand, tended to increase recidivism slightly.

Patrick H. Tolan of the University of Illinois at Chicago has also recently published an empirical review of delinquency programs. To Lipsey's findings he adds that "family interventions have repeatedly shown efficacy for reducing antisocial behavior and appear to be among the most promising interventions to date." According to Forgatch, two experiments in Eugene, Oregon, showed that teaching parents better monitoring and more consistent, less coercive discipline techniques reduces their kids' misbehavior. "We should make parenting skills classes compulsory for high school students," argues Raine of the University of Southern California.

Unfortunately, Tolan observes, family intervention is difficult and rarely attempted. The most common kinds of programs—counseling by social workers, peer mediation, and neighborhood antiviolence initiatives—are hardly ever examined to see whether they produce lasting benefits. "It usually is hard to imagine that a good idea put into action by well-meaning and enlightened people cannot help," he noted in the paper. "It may seem that any effort is better than nothing. Yet our review and several of the more long-term and sophisticated analyses suggest that both of these assumptions can be dangerously wrong. Not only have programs that have been earnestly launched been ineffective, but some of our seemingly best ideas have led to worsening of the behavior of those subjected to the intervention."

Many researchers are thus frustrated that the Violent Crime Control and Law Enforcement Act of 1994 puts most of its $6.1 billion for crime prevention in untested and controversial programs, such as "midnight basketball" and other after-school activities. "Maybe these programs will help; maybe they won't," Tolan says. "No one has done a careful evaluation." The Crime Act does not insist that grant applicants demonstrate or even measure the effectiveness of their approach. For these and other reasons, Republicans vowed in their 1994 "Contract with America" to repeal all prevention programs in the Crime Act and to increase funding for prison construction. But that strategy ignored research. "We do know," Tolan asserts, "that locking kids up will not reduce crime and may eventually make the problem worse."

The failure of sociology to demonstrate conclusively effective means of controlling violent crime has made some impatient. "There is a growing recognition that we're not going to solve any problem in society using just one discipline," says Diana Fishbein, a professor of criminology at the University of Baltimore. "Sociological factors play a role. But they have not been able to explain why one person becomes violent and another doesn't."

Some social scientists are looking to psychiatrists, neurologists, and geneticists to provide answers to that question, ready or not. "Science must tell us what individuals will or will not become criminals, what individuals will or will not become victims, and what law enforcement strategies will or will not work," wrote C. Ray Jeffery, a criminologist at Florida State University, in 1994 in the *Journal of Research in Crime and Delinquency*.

■ BIOLOGICAL FACTORS

As medical researchers have teased out a few tantalizing links between brain chemistry, heredity, hormones, physiology, and assaultive behavior, some have become emboldened. "Research in the past 10 years conclusively demonstrates that biological factors play some role in the etiology of violence. That is scientifically beyond doubt," Raine holds forth. The importance of that role is still very much in doubt, however.

As with social risk factors, no biological abnormality has been shown to *cause* violent aggression—nor is that likely except in cases of extreme psychiatric disorder. But researchers have spotted several unusual features, too subtle even to be considered medical problems, that tend to appear in the bodies and brains of physically aggressive men. On average, for example, they have higher levels of testosterone, a sex hormone important for building muscle mass and strength, among other functions. James M. Dabbs, Jr., of Georgia State University has found in his experiments with prison inmates that men with the highest testosterone concentrations are more likely to have committed violent crimes. But Dabbs emphasizes that the link is indirect and "mediated by many social factors," such as higher rates of divorce and substance abuse.

"Low resting heart rate probably represents the best replicated biological correlate of antisocial behavior," Raine observes, pointing to 14 studies that have found that problem children and petty criminals tend to have significantly lower pulses than do well-behaved counterparts. A slower heartbeat "probably reflects fearlessness and underarousal," Raise theorizes. "If we lack the fear of getting hurt, it may lead to a predisposition to engage in violence." But that hypothesis fails to explain why at least 15 studies have failed to find abnormal heart rates in psychopaths.

Jerome Kagan, a Harvard University psychologist, has suggested that an inhibited "temperament" may explain why the great majority of children from high-risk homes grow up to become law-abiding citizens. One study tested pulse, pupil dilation, vocal tension, and blood levels of the neurotransmitter norepinephrine and the stress-regulating hormone cortisol to distinguish inhibited from uninhibited, underaroused two-year-olds. An expert panel on "Understanding and Preventing Violence" convened by the National Research Council suggested in its 1993 report that inhibited children may be protected by their fearfulness from becoming aggressive, whereas uninhibited children may be prone to later violence. The panel concluded that "although such factors in isolation may not be expected to be strong predictors of violence, in conjunction with other early family and cognitive measures, the degree of prediction may be considerable."

Perhaps the most frequently cited biological correlate of violent behavior is a low level of serotonin, a chemical that in the body inhibits the secretion of stomach acid and stimulates smooth muscle and in the brain functions as a neurotransmitter. A large body of animal evidence links low levels of serotonin to impulsive aggression. Its role in humans is often oversimplified, however. "Serotonin has a calming effect on behavior by reducing the level of violence," Jeffery wrote in 1993 in the *Journal of Criminal Justice Education*. "Thus, by increasing the level of serotonin in the brain, we can reduce the level of violence." A front-page article in December 1993 in the

Chicago Tribune explained that "when serotonin declines . . . impulsive aggression is unleashed."

Such explanations do violence to the science. In human experiments, researchers do not generally have access to the serotonin inside their subject's braincase. Instead they tap cerebrospinal fluid from the spinal column and measure the concentration of 5-hydroxyindoleacetic acid (5-HIAA), which is produced when serotonin is used up and broken down by the enzyme monoamine oxidase (MAO). Serotonin does its job by binding to any of more than a dozen different neural receptors, each of which seems to perform a distinct function. The low levels of 5-HIAA seen in violent offenders may indicate a shortage of serotonin in the brain or simply a dearth of MAO—in which case their serotonin levels may actually be high. Moreover, serotonin can rise or drop in different regions of the brain at different times, with markedly different effects.

Environment, too, plays a role: nonhuman primate studies show that serotonin often fluctuates with pecking order, dropping in animals when they are threatened and rising when they assume a dominant status. The numerous pathways through which serotonin can influence mood and behavior confound attempts to simply "reduce the level of violence" by administering serotonin boosters such as Prozac, a widely prescribed antidepressant. Nevertheless, the link between 5-HIAA and impulsive aggression has led to a concerted hunt for the genes that control the production and activity of serotonin and several other neurotransmitters. "Right now we have in our hand many of the genes that affect brain function," says David Goldman, chief of neurogenetics at the National Institute on Alcohol Abuse and Alcoholism. Although none has yet been shown to presage violence, "I believe the markers are there," he says. But he warns that "we're going to have to understand a whole lot more about the genetic, environmental, and developmental origins of personality and psychiatric disease" before making use of the knowledge.

Yudofsky is less circumspect. "We are on the verge of a revolution in genetic medicine," he asserts. "The future will be to understand the genetics of aggressive disorders and to identify those who have greater tendencies to become violent."

Few researchers believe genetics alone will ever yield reliable predictors of behavior as complex and multifarious as harmful aggression. Still, the notion that biologists and sociologists might together be able to assemble a complicated model that can scientifically pick out those who pose the greatest threat of vicious attack seems to be gaining currency. Already some well-respected behavioral scientists are advocating a medical approach to crime control based on screening, diagnostic prediction, and treatment. "A future generation will reconceptualize nontrivial recidivistic crime as a disorder," Raine predicted in his book, *The Psychopathology of Crime.*

■ COMPULSORY TREATMENT?

But the medical model for crime may be fraught with peril. When the "disease" is intolerable behavior that threatens society, will "treatment" necessarily be compulsory and indefinite? If, to reexamine Raine's hypothetical example, prediction models are judged as reliable but "biological, social, and cognitive intervention programs" are not, might eight-year-old boys be judged incorrigible before they have broken any law? Calls for screening are now heard more often. "There are areas where we can begin to incorporate biological approaches," Fishbein argues. "Delinquents need to be individually assessed." Masters claims that "we now know enough about the serotonergic system so that if we see a kid doing poorly in school, we ought to look at his serotonin levels."

In his article Jeffery emphasized that "attention must focus on the 5 percent of the delinquent population who commit 50 percent of the offenses. . . . This effort must identify high-risk persons at an early age and place them in treatment programs before they have committed the 10 to 20 major felonies characteristic of the career criminal."

Yudofsky suggests a concrete method to do this: "You could ask parents whether they consider their infant high-strung or hyperactive. Then screen more closely by challenging the infants with provocative situations." When kids respond too aggressively, he suggests "you could do careful neurologic testing and train the family how not to goad and fight them. Teach the children nonviolent ways to reduce frustration. And when these things don't work, consider medical interventions, such as beta blockers, anticonvulsants, or lithium."

"We haven't done this research, but I have no doubt that it would make an enormous impact and would be immediately cost-effective," Yudofsky continues. While he bemoans a lack of drugs designed specifically to treat aggression, he sees a tremendous "opportunity for the pharmaceutical industry," which he maintains is "finally getting interested."

But some worry that voluntary screening for the good of the child might lead to mandatory screening for the protection of society. "It is one thing to convict someone of an offense and compel them to do something. It is another thing to go to someone who has not done anything wrong and say, 'You look like a high risk, so you have to do this,'" Akers observes. "There is a very clear ethical difference, but that is a very thin line that people, especially politicians, might cross over."

Even compelling convicted criminals to undergo treatment raises thorny ethical issues. Today the standards for proving that an offender is so mentally ill that he poses a danger to himself or others and thus can be incarcerated indefinitely are quite high. The medical model of violent crime threatens to lower those standards substantially. Indeed, Jeffery argues that "if we are to follow the medical model, we must use neurological examinations in place of the insanity defense and the concept of guilt. Criminals must be placed in medical clinics, not prisons." Fishbein says she is "beginning to think that treatment should be mandatory. We don't ask offenders whether they want to be incarcerated or executed. They should remain in a secure facility until they can show without a doubt that they are self-controlled." And if no effective treatments are available? "They should be held indefinitely," she says.

■ MORAL IMPERATIVE

Unraveling the mystery of human behavior, just like untangling the human genetic code, creates a moral imperative to use that knowledge. To ignore it—to imprison without treatment those whom society defines as sick for the behavioral symptoms of their illness—is morally indefensible. But to replace a fixed term of punishment set by the conscience of a society with forced therapy based on the judgment of scientific experts is to invite even greater injustice.

—March 1995

V

Disease of the Brain and Disorder of the Mind

ATTENTION-DEFICIT HYPERACTIVITY DISORDER

A new theory suggests the disorder results from a failure in self-control. ADHD may arise when key brain circuits do not develop properly, perhaps because of an altered gene or genes

Russell A. Barkley

As I watched five-year-old Keith in the waiting room of my office, I could see why his parents said he was having such a tough time in kindergarten. He hopped from chair to chair, swinging his arms and legs restlessly, and then began to fiddle with the light switches, turning the lights on and off again to everyone's annoyance—all the while talking nonstop. When his mother encouraged him to join a group of other children busy in the playroom, Keith butted into a game that was already in progress and took over, causing the other children to complain of his bossiness and drift away to other activities. Even when Keith had the toys to himself, he fidgeted aimlessly with them and seemed unable to entertain himself quietly. Once I examined him more fully, my initial suspicions were confirmed: Keith had attention-deficit hyperactivity disorder (ADHD).

Since the 1940s, psychiatrists have applied various labels to children who are hyperactive and inordinately inattentive and impulsive. Such youngsters have been considered to have "minimal brain dysfunction," "brain-injured child syndrome," "hyperkinetic reaction of childhood," "hyperactive child syndrome," and, most recently, "attention-deficit disorder." The frequent name changes reflect how uncertain researchers have been about the underlying causes of, and even the precise diagnostic criteria for, the disorder.

Within the past several years, however, those of us who study ADHD have begun to clarify its symptoms and causes and have found that it may have a genetic underpinning. Today's view of the basis of the condition is strikingly different from that of just a few years ago. We are finding that ADHD is not a disorder of attention per se, as had long been assumed. Rather it arises as a developmental failure in the brain circuitry that underlies inhibition and self-control. This loss of self-control in turn impairs other important brain functions crucial for maintaining attention, including the ability to defer immediate rewards for later, greater gain.

ADHD involves two sets of symptoms: inattention and a combination of hyperactive and impulsive behaviors [*see box on page 221*]. Most children are more active, distractible, and impulsive than adults. And they are more inconsistent, affected by momentary events and dominated by objects in their immediate environment. The younger the children, the less able they are to be aware of time or to give priority to future events over more immediate wants. Such behaviors are signs of a problem, however, when children display them significantly more than their peers do.

Boys are at least three times as likely as girls to develop the disorder; indeed, some studies have found that boys with ADHD outnumber girls with the condition by nine to one, possibly because boys are genetically more prone to disorders of the nervous system. The behavior patterns that typify ADHD usually arise between the ages of three and five. Even so, the age of onset can vary widely: some children do not develop symptoms until late childhood or even early adolescence. Why their symptoms are delayed remains unclear.

Huge numbers of people are affected. Many studies estimate that between 2 and 9.5 percent of all school-age children worldwide have ADHD; researchers have identified it in every nation and culture they have studied. What is more, the condition, which was once thought to ease with age, can persist into adulthood. For example, roughly two thirds of 158 children with ADHD my colleagues and I evaluated in the 1970s still had the disorder in their twenties. And many of those who no

DIAGNOSING ADHD

Psychiatrists diagnose attention-deficit hyperactivity disorder (ADHD) if the individual displays six or more of the following symptoms of inattention or six or more symptoms of hyperactivity and impulsivity. The signs must occur often and be present for at least six months to a degree that is maladaptive and inconsistent with the person's developmental level. In addition, some of the symptoms must have caused impairment before the age of seven and must now be causing impairment in two or more settings. Some must also be leading to significant impairment in social, academic, or occupational functioning; none should occur exclusively as part of another disorder. (Adapted with permission from the fourth edition of the *Diagnostic and Statistical Manual of Mental Disorders,* © 1994 American Psychiatric Association.)

Inattention

- Fails to give close attention to details or makes careless mistakes in schoolwork, work, or other activities
- Has difficulty sustaining attention in tasks or play activities
- Does not seem to listen when spoken to directly
- Does not follow through on instructions and fails to finish schoolwork, chores, or duties in the workplace
- Has difficulty organizing tasks and activities
- Avoids, dislikes, or is reluctant to engage in tasks that require sustained mental effort (such as schoolwork)
- Loses things necessary for tasks or activities (such as toys, school assignments, pencils, books, or tools)
- Is easily distracted by extraneous stimuli
- Is forgetful in daily activities

Hyperactivity and Impulsivity

- Fidgets with hands or feet or squirms in seat
- Leaves seat in classroom or in other situations in which remaining seated is expected
- Runs about or climbs excessively in situations in which it is inappropriate (in adolescents or adults, subjective feelings of restlessness)
- Has difficulty playing or engaging in leisure activities quietly
- Is "on the go" or acts as if "driven by a motor"
- Talks excessively
- Blurts out answers before questions have been completed
- Has difficulty awaiting turns
- Interrupts or intrudes on others

longer fit the clinical description of ADHD were still having significant adjustment problems at work, in school, or in other social settings.

To help children (and adults) with ADHD, psychiatrists and psychologists must better understand the causes of the disorder. Because researchers have traditionally viewed ADHD as a problem in the realm of attention, some have suggested that it stems from an inability of the brain to filter competing sensory inputs, such as sights and sounds. But recently scientists led by Joseph A. Sergeant of the University of Amsterdam have shown that children with ADHD do not have difficulty in that area; instead they cannot inhibit their impulsive motor responses to such input. Other researchers have found that children with ADHD are less capable of preparing motor responses in anticipation of events and are insensitive to feedback about errors made in those responses. For example, in a commonly used test of reaction time, children with ADHD are less able than other children to ready themselves to press one of several keys when they see a warning light. They also do not slow down after making mistakes in such tests in order to improve their accuracy.

■ THE SEARCH FOR A CAUSE

No one knows the direct and immediate causes of the difficulties experienced by children with ADHD, although advances in neurological imaging techniques and genetics promise to clarify this issue over the next five years. Already they have yielded clues, albeit ones that do not yet fit together into a coherent picture.

Imaging studies over the past decade have indicated which brain regions might malfunction in patients with ADHD and thus account for the symptoms of the condition. That work suggests the involvement of the prefrontal cortex, part of the cerebellum, and at least two of the clusters of nerve cells deep in the brain that are collectively known as the basal ganglia [see color plate 3]. In a 1996 study F. Xavier Castellanos, Judith L. Rapoport, and their colleagues at the National Institute of Mental Health found that the right prefrontal cortex and two basal ganglia called the caudate nucleus and the globus pallidus are significantly smaller than normal in children with ADHD. Earlier this year Castellanos's group found that the vermis region of the cerebellum is also smaller in ADHD children.

The imaging findings make sense because the brain areas that are reduced in size in children with ADHD are the very ones that regulate attention. The right prefrontal cortex, for example, is involved in "editing" one's behavior, resisting distractions, and developing an awareness of self and time. The caudate nucleus and the globus pallidus help to switch off automatic responses to allow more careful deliberation by the cortex and to coordinate neurological input among various regions of the cortex. The exact role of the cerebellar vermis is unclear, but early studies suggest it may play a role in regulating motivation.

What causes these structures to shrink in the brains of those with ADHD? No one knows, but many studies have suggested that mutations in several genes that are normally very active in the prefrontal cortex and basal ganglia might play a role. Most researchers now believe that ADHD is a polygenic disorder—that is, that more than one gene contributes to it.

Early tips that faulty genetics underlie ADHD came from studies of the relatives of children with the disorder. For instance, the siblings of children with ADHD are between five and seven times more likely to develop the syndrome than children from unaffected families. And the children of a parent who has ADHD have up to a 50 percent chance of experiencing the same difficulties.

The most conclusive evidence that genetics can contribute to ADHD, however, comes from studies of twins. Jacquelyn J. Gillis, then at the University of Colorado, and her colleagues reported in 1992 that the ADHD risk of a child whose identical twin has the disorder is between 11 and 18 times greater than that of a nontwin sibling of a child with ADHD; between 55 and 92 percent of the identical twins of children with ADHD eventually develop the condition.

One of the largest twin studies of ADHD was conducted by Helene Gjone and Jon M. Sundet of the University of Southampton in England. It involved 526 identical twins, who inherit exactly the same genes, and 389 fraternal twins, who are no more alike genetically than siblings born years apart. The team found that ADHD has a heritability approaching 80 percent, meaning that up to 80 percent of the differences in attention, hyperactivity, and impulsivity between people with ADHD and those without the disorder can be explained by genetic factors.

Nongenetic factors that have been linked to ADHD include premature birth, maternal alcohol and tobacco use, exposure to high levels of lead in early childhood, and brain injuries—especially those that involve the prefrontal cortex. But even together, these factors can account for only between 20 and 30 percent of ADHD cases among boys; among girls, they account for an even smaller percentage. (Contrary to popular belief, neither dietary factors, such as the amount of sugar a child consumes, nor poor child-rearing methods have been consistently shown to contribute to ADHD.)

Which genes are defective? Perhaps those that dictate the way in which the brain uses dopamine, one of the chemicals known as neurotransmitters that convey messages from one nerve cell, or neuron, to another. Dopamine is secreted by neurons in specific parts of the brain to inhibit or modulate the activity of other neurons, particularly those involved in emotion and movement. The movement disorders of Parkinson's disease, for example, are caused by the death of dopamine-secreting neurons in a region of the brain underneath the basal ganglia called the substantia nigra.

Some impressive studies specifically implicate genes that encode, or serve as the blueprint for, dopamine receptors and transporters; these genes are very active in the prefrontal cortex and basal ganglia. Dopamine receptors sit on the surface of certain neurons. Dopamine delivers its message to those neurons by binding to the receptors. Dopamine transporters protrude from neurons that secrete the neurotransmitter; they take up unused dopamine so that it can be used again. Mutations in the dopamine receptor gene can render receptors less sensitive to dopamine. Conversely, mutations in the dopamine transporter gene can yield overly effective transporters that scavenge secreted dopamine before it has a chance to bind to dopamine receptors on a neighboring neuron.

In 1995 Edwin H. Cook and his colleagues at the University of Chicago reported that children with ADHD were more likely than others to have a particular variation in the dopamine transporter gene *DAT1*. Similarly, in 1996 Gerald J. LaHoste of the University of California at Irvine and his co-workers found that a variant of the dopamine receptor gene *D4* is more common among children with ADHD. But each of these studies involved 40 or 50 children—a relatively small number—so their findings are now being confirmed in larger studies.

■ FROM GENES TO BEHAVIOR

How do the brain-structure and genetic defects observed in children with ADHD lead to the characteristic behaviors of the disorder? Ultimately, they might be found to underlie impaired behavioral inhibition and self-control, which I have concluded are the central deficits in ADHD.

Self-control—or the capacity to inhibit or delay one's initial motor (and perhaps emotional) responses to an event—is a critical foundation for the performance of any task. As most children grow up, they gain the ability to engage in mental activities, known as executive functions, that help them deflect distractions, recall goals, and take the steps needed to reach them. To achieve a goal in work or play, for instance, people need to be able to remember their aim (use hindsight), prompt themselves about what they need to do to reach that goal (use forethought), keep their emotions reined in, and motivate themselves. Unless a person can inhibit interfering thoughts and impulses, none of these functions can be carried out successfully.

In the early years, the executive functions are performed externally: children might talk out loud to themselves while remembering a task or puzzling out a problem. As children mature, they internalize, or make private, such executive functions, which prevents others from knowing their thoughts. Children with ADHD, in contrast, seem to lack the restraint needed to inhibit the public performance of these executive functions.

The executive functions can be grouped into four mental activities. One is the operation of working memory—holding information in the mind while working on a task, even if the original stimulus that provided the information is gone. Such remembering is crucial to timeliness and goal-directed behavior: it provides the means for hindsight, forethought, preparation, and the ability to imitate the complex, novel behavior of others—all of which are impaired in people with ADHD.

The internalization of self-directed speech is another executive function. Before the age of six, most children speak out loud to themselves frequently, reminding themselves how to perform a particular task or trying to cope with a problem, for example. ("Where did I put that book? Oh, I left it under the desk.") In elementary school, such private speech evolves into inaudible muttering; it usually disappears by age 10. Internalized, self-directed speech allows one to reflect to oneself, to follow rules and instructions, to use self-questioning as a form of problem solving, and to construct "meta-rules," the basis for understanding the rules for using rules—all quickly and without tipping one's hand to others. Laura E. Berk and her colleagues at Illinois State University reported in 1991 that the internalization of self-directed speech is delayed in boys with ADHD.

A third executive mental function consists of controlling emotions, motivation, and state of arousal. Such control helps individuals achieve goals by enabling them to delay or alter potentially distracting emotional reactions to a particular event and to generate private emotions and motivation. Those who rein in their immediate passions can also behave in more socially acceptable ways.

The final executive function, reconstitution, actually encompasses two separate processes: breaking down observed behaviors and combining the parts into new actions not previously learned from experience. The capacity for reconstitution gives humans a great degree of fluency, flexibility, and creativity; it allows individuals to propel themselves toward a goal without having to learn all the needed steps by rote. It permits children as they mature to direct their behavior across increasingly longer intervals by combining behaviors into ever longer chains to attain a goal. Initial studies imply that children with ADHD are less capable of reconstitution than are other children.

I suggest that like self-directed speech, the other three executive functions become internalized during typical neural development in early childhood. Such privatization is essential for creating visual imagery and verbal thought. As children grow up, they develop the capacity to behave covertly, to mask some of their behaviors or feelings from others. Perhaps because of faulty genetics or embryonic development, children with ADHD have not attained this ability and therefore display too much public behavior and speech. It is my assertion that the inattention, hyperactivity, and impulsivity of children with ADHD are caused by their failure to be guided by internal instruction and by their inability to curb their own inappropriate behavior.

A PSYCHOLOGICAL MODEL OF ADHD

A loss of behavioral inhibition and self-control leads to the following disruptions in brain functioning:

Impaired Function	Consequence	Example
Nonverbal working memory	Diminished sense of time Inability to hold events in mind Defective hindsight Defective forethought	Nine-year-old Jeff routinely forgets important responsibilities, such as deadlines for book reports or an after-school appointment with the principal
Internalization of self-directed speech	Deficient rule-governed behavior Poor self-guidance and self-questioning	Five-year-old Audrey talks too much and cannot give herself useful directions silently on how to perform a task
Self-regulation of mood, motivation, and level of arousal	Displays all emotions publicly; cannot censor them Diminished self-regulation of drive and motivation	Eight-year-old Adam cannot maintain the persistent effort required to read a story appropriate for his age level and is quick to display his anger when frustrated by assigned schoolwork
Reconstitution (ability to break down observed behaviors into component parts that can be recombined into new behaviors in pursuit of a goal)	Limited ability to analyze behaviors and synthesize new behaviors Inability to solve problems	Fourteen-year-old Ben stops doing a homework assignment when he realizes that he has only two of the five assigned questions; he does not think of a way to solve the problem, such as calling a friend to get the other three questions

■ PRESCRIBING SELF-CONTROL

If, as I have outlined, ADHD is a failure of behavioral inhibition that delays the ability to privatize and execute the four executive mental functions I have described, the finding supports the theory that children with ADHD might be helped by a more structured environment. Greater structure can be an important complement to any drug therapy the children might receive. Currently children (and adults) with ADHD often receive drugs such as Ritalin that boost their capacity to inhibit and regulate impulsive behaviors. These drugs act by inhibiting the dopamine transporter, increasing the time that dopamine has to bind to its receptors on other neurons.

Such compounds (which, despite their inhibitory effects, are known as psychostimulants) have been found to improve the behavior of between 70 and 90 percent of children with ADHD older than five years. Children with ADHD who take such medication not only are less impulsive, restless, and distractible but are also better able to hold important information in mind, to be more productive academically, and to have more internalized speech and better self-control. As a result, they tend to be liked better by other children and to experience less punishment for their actions, which improves their self-image.

My model suggests that in addition to psychostimulants—and perhaps antidepressants, for some children—treatment for ADHD should include training parents and teachers in specific and more effective methods for managing the behavioral problems of children with the disorder. Such methods involve making the consequences of a child's actions more frequent and immediate and increasing the external use of prompts and cues about rules and time intervals. Parents and teachers must aid children with ADHD by anticipating events for them, breaking future tasks down into smaller and more immediate steps, and using artificial immediate rewards. All these steps serve to externalize time, rules, and consequences as a replacement for the weak internal forms of information, rules, and motivation of children with ADHD.

In some instances, the problems of ADHD children may be severe enough to warrant their placement in special education programs. Although such programs are not intended as a cure for the child's difficulties, they typically do provide a smaller, less competitive, and more supportive environment in which the child can receive individual instruction. The hope is that once children learn techniques to overcome their deficits in self-control, they will be able to function outside such programs.

There is no cure for ADHD, but much more is now known about effectively coping with and managing this persistent and troubling developmental disorder. The day is not far off when genetic testing for ADHD may become available and more specialized medications may be designed to counter the specific genetic deficits of the children who suffer from it.

—September 1998

AUTISM

Autistic individuals suffer from a biological defect.
Although they cannot be cured, much can be done
to make life more hospitable for them

Uta Frith

The image often invoked to describe autism is that of a beautiful child imprisoned in a glass shell. For decades, many parents have clung to this view, hoping that one day a means might be found to break the invisible barrier. Cures have been proclaimed, but not one of them has been backed by evidence. The shell remains intact. Perhaps the time has come for the whole image to be shattered. Then at last we might be able to catch a glimpse of what the minds of autistic individuals are truly like.

Psychological and physiological research has shown that autistic people are not living in rich inner worlds but instead are victims of a biological defect that makes their minds very different from those of normal individuals. Happily, however, autistic people are not beyond the reach of emotional contact.

Thus, we can make the world more hospitable for autistic individuals just as we can, say, for the blind. To do so, we need to understand what autism is like—a most challenging task. We can imagine being blind, but autism seems unfathomable. For centuries, we have known that blindness is often a peripheral defect at the sensory-motor level of the nervous system, but only recently has autism been appreciated as a central defect at the highest level of cognitive processing. Autism, like blindness, persists throughout life, and it responds to special efforts in compensatory education. It can give rise to triumphant feats of coping but can also lead to disastrous secondary consequences—anxiety, panic, and depression. Much can be done to prevent problems. Understanding the nature of the handicap must be the first step in any such effort.

Autism existed long before it was described and named by Leo Kanner of the Johns Hopkins Children's Psychiatric Clinic. Kanner published his landmark paper in 1943 after he had observed 11 children who seemed to him to form a recognizable group. All had in common four traits: a preference for aloneness, an insistence on sameness, a liking for elaborate routines, and some abilities that seemed remarkable compared with the deficits.

Concurrently, though quite independently, Hans Asperger of the University Pediatric Clinic in Vienna prepared his doctoral thesis on the same type of child. He also used the term "autism" to refer to the core features of the disorder. Both men borrowed the label from adult psychiatry, where it had been used to refer to the progressive loss of contact with the outside world experienced by schizophrenics. Autistic children seemed to suffer such a lack of contact with the world around them from a very early age.

Kanner's first case, Donald, has long served as a prototype for diagnosis. It had been evident early in life that the boy was different from other children. At two years of age, he could hum and sing tunes accurately from memory. Soon he learned to count to 100 and to recite both the alphabet and the 25 questions and answers of the Presbyterian catechism. Yet he had a mania for making toys and other objects spin. Instead of playing like other toddlers, he arranged beads and other things in groups of different colors or threw them on the floor, delighting in the sounds they made. Words for him had a literal, inflexible meaning.

Donald was first seen by Kanner at age five. Kanner observed that the boy paid no attention to people around him. When someone interfered with his solitary activities, he was never angry with the interfering person but impatiently removed the hand that was in his way. His mother was the only person with whom he had any significant contact, and that seemed attributable mainly to the great effort she made to share activities with him. By the time Donald was about eight years old, his conversation consisted largely of repetitive questions. His relation to people remained limited to his immediate wants and needs, and his attempts at contact stopped as soon as he was told or given what he had asked for.

Some of the other children Kanner described were mute, and he found that even those who spoke did not really communicate but used language in a very odd way. For example, Paul, who was five, would parrot speech verbatim. He would say "You want candy" when he meant "I want candy." He was in the habit of repeating, almost every day, "Don't throw the dog off the balcony," an utterance his mother traced to an earlier incident with a toy dog.

Twenty years after he had first seen them, Kanner reassessed the members of his original group of children. Some of them seemed to have adapted socially much better than others, although their failure to communicate and to form relationships remained, as did their pedantry and single-mindedness. Two prerequisites for better adjustment, though no guarantees of it, were the presence of speech before age five and relatively high intellectual ability. The brightest autistic individuals had, in their teens, become uneasily aware of their peculiarities and had made conscious efforts to conform. Nevertheless, even the best adapted were rarely able to be self-reliant or to form friendships. The one circumstance that seemed to be helpful in all the cases was an extremely structured environment.

As soon as the work of the pioneers became known, every major clinic began to identify autistic children. It was found that such children, in addition to their social impairments, have substantial intellectual handicaps. Although many of them perform relatively well on certain tests, such as copying mosaic patterns with blocks, even the most able tend to do badly on test questions that can be answered only by the application of common sense.

Autism is rare. According to the strict criteria applied by Kanner, it appears in four of every 10,000 births. With the somewhat wider criteria used in current diagnostic practice, the incidence is much higher: one or two in 1,000 births, about the same as Down's syndrome. Two to four times as many boys as girls are affected.

For many years, autism was thought to be a purely psychological disorder without an organic basis. At first, no obvious neurological problems were found. The autistic children did not necessarily have low intellectual ability, and they often looked physically normal. For these reasons, psychogenic theories were proposed and taken seriously for many years. They focused on the idea that a child could become autistic because of some existentially threatening experience. A lack of maternal bonding or a disastrous experience of rejection, so the theory went, might drive an infant to withdraw into an inner world of fantasy that the outside world never penetrates.

These theories are unsupported by any empirical evidence. They are unlikely to be supported because there are many instances of extreme rejection and deprivation in childhood, none of which have resulted in autism. Unfortunately, therapies vaguely based on such notions are still putting pressure on parents to accept a burden of guilt for the supposedly avoidable and reversible breakdown of interpersonal interactions. In contrast, well-structured behavior modification programs have often helped families in the management of autistic children, especially children with severe behavior problems. Such programs do not claim to reinstate normal development.

The insupportability of the psychogenic explanation of autism led a number of workers to search for a biological cause. Their efforts implicate a defective structure in the brain, but that structure has not yet been identified. The defect is believed to affect the thinking of autistic people, making them unable to evaluate their own thoughts or to perceive clearly what might be going on in someone else's mind.

Autism appears to be closely associated with several other clinical and medical conditions. They include maternal rubella and chromosomal abnormality, as well as early injury to the brain and infantile seizures. Most impressive, perhaps, are studies showing that autism can have a genetic basis. Both identical twins are much more likely to be autistic than are both fraternal twins. Moreover, the likelihood that autism will occur twice in the same family is 50 to 100 times greater than would be expected by chance alone.

Structural abnormalities in the brains of autistic individuals have turned up in anatomic studies and brain-imaging procedures. Both epidemiological and neuropsychological studies have demonstrated that autism is strongly correlated with mental retardation, which is itself clearly linked to physiological abnormality. This fact fits well with the idea that autism results from a distinct brain abnormality that is often part of more extensive damage. If the abnormality is pervasive, the mental retardation will be more severe, and the likelihood of damage to the critical brain system will increase. Conversely, it is possible for the critical system alone to be damaged. In such cases, autism is not accompanied by mental retardation.

Neuropsychological testing has also contributed evidence for the existence of a fairly circumscribed brain abnormality. Autistic individuals who are otherwise able show specific and extensive deficits on certain tests that involve planning, initiative, and spontaneous generation of new ideas. The same deficits appear in patients who have frontal lobe lesions. Therefore, it seems plausible that whatever the defective brain structure is, the frontal lobes are implicated.

Population studies carried out by Lorna Wing and her colleagues at the Medical Research Council's Social Psychiatry Unit in London reveal that the different symptoms of autism do not occur together simply by coincidence. Three core features in particular—impairments in communication, imagination, and socialization—form a distinct triad. The impairment in communication includes such diverse phenomena as muteness and delay in learning to talk, as well as problems in

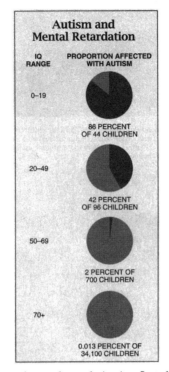

Autism and Mental Retardation

IQ RANGE / **PROPORTION AFFECTED WITH AUTISM**

0–19 — 86 PERCENT OF 44 CHILDREN

20–49 — 42 PERCENT OF 96 CHILDREN

50–69 — 2 PERCENT OF 700 CHILDREN

70+ — 0.013 PERCENT OF 34,100 CHILDREN

Close link between autism and mental retardation is reflected in this chart. The percentage of children showing the social impairments typical of autism is highest at low levels of intelligence as measured by tests in which an intelligence quotient (IQ) below 70 is subnormal. For example, 86 percent of 44 children in the lowest IQ range showed the social impairments of autism. The data are drawn from a population of about 35,000 children aged under 15 years. (Lorna Wing, Medical Research Council, London)

comprehending or using nonverbal body language. Other autistic individuals speak fluently but are overliteral in their understanding of language. The impairment in imagination appears in young autistic children as repetitive play with objects and in some autistic adults as an obsessive interest in facts. The impairment in socialization includes ineptness and inappropriate behavior in a wide range of reciprocal social interactions, such as the ability to make and keep friends. Nevertheless, many autistic individuals prefer to have company and are eager to please.

The question is why these impairments, and only these, occur together. The challenge to psychological theorists was clear: to search for a single cognitive component that would explain the deficits yet still allow for the abilities that autistic people display in certain aspects of interpersonal interactions. My colleagues at the Medical Research Council's Cognitive Development Unit in London and I think we have identified just such a component. It is a cognitive mechanism of a highly complex and abstract nature that could be described in computational terms. As a shorthand, one can refer to this component by one of its main functions, namely the ability to think about thoughts or to imagine another individual's state of mind. We propose that this component is damaged in autism. Furthermore, we suggest that this mental component is innate and has a unique brain substrate. If it were possible to pinpoint that substrate—whether it is an anatomic structure, a physiological system, or a chemical pathway—one might be able to identify the biological origin of autism.

The power of this component in normal development becomes obvious very early. From the end of the first year onward, infants begin to participate in what has been called shared attention. For example, a normal child will point to something for no reason other than to share his interest in it with someone else. Autistic children do not show shared attention. Indeed, the absence of this behavior may well be one of the earliest signs of autism. When an autistic child points at an object, it is only because he wants it.

In the second year of life, a particularly dramatic manifestation of the critical component can be seen in normal children: the emergence of pretense, or the ability to engage in fantasy and pretend play. Autistic children cannot understand pretense and do not pretend when they are playing. The difference can be seen in such a typical nursery game as "feeding" a teddy bear or a doll with an empty spoon. The normal child goes through the appropriate motions of feeding and accompanies the action with appropriate slurping noises. The autistic child merely twiddles or flicks the spoon repetitively. It is precisely the absence of early and simple communicative behaviors, such as shared attention and make-believe play, that often creates the first nagging doubts in the minds of the parents about the development of their child. They rightly feel that they cannot engage the child in the emotional to-and-fro of ordinary life.

My colleague Alan M. Leslie devised a theoretical model of the cognitive mechanisms underling the key abilities of shared attention and pretense. He postulates an innate mechanisms whose function is to form and use what we might call second-order representations. The world around us consists not only of visible bodies and events, captured by first-order representations, but also of invisible minds and mental events, which require second-order representation. Both types of representation have to be kept in mind and kept separate from each other.

Second-order representations serve to make sense of otherwise contradictory or incongruous information. Suppose a normal child, Beth, sees her mother holding a banana in such a way as to be pretending that it is a telephone. Beth has in mind facts about bananas and facts about telephones—first-order representations. Nevertheless, Beth is not the least bit confused and will not start eating telephones or talking to bananas. Confusion is avoided because Beth computes from the concept of pretending (a second-order representation) that her mother is engaging simultaneously in an imaginary activity and a real one.

As Leslie describes the mental process, pretending should be understood as computing a three-term relation between an actual situation, an imaginary situation, and an agent who does the pretending. The imaginary situation is then not treated as the real situation. Believing can be understood in the same way as pretending. This insight enabled us to predict that autistic children, despite an adequate mental age (above four years or so), would not be able to understand that someone can have a mistaken belief about the world.

Together with our colleague Simon Baron-Cohen, we tested this prediction by adapting an experiment originally devised by two Australian developmental psychologists, Heinz Wimmer and Josef Perner. The test has become known as the Sally-Anne task. Sally and Anne are playing together. Sally has a marble that she puts in a basket before leaving the room. While she is out, Anne moves the marble to a box. When Sally returns, wanting to retrieve the marble, she of course looks in the basket. If this scenario is presented as, say, a puppet show to normal children who are four years of age or more, they understand that Sally will look in the basket even though they know the marble is not there. In other words, they can represent Sally's erroneous belief as well as the true state of things. Yet in our test, 16 of 20 autistic children with a mean mental age of nine failed the task—answering that Sally would look in the box—in spite of being able to answer correctly a variety of other questions relating to the facts of the episode. They could not conceptualize the possibility that Sally believed something that was not true.

Many comparable experiments have been carried out in other laboratories, which have largely confirmed our prediction: autistic children are specifically impaired in their understanding of mental states. They appear to lack the innate component underlying this ability. This component, when it works normally, has the most far-reaching consequences for higher-order conscious processes. It underpins the special feature of the human mind, the ability to reflect on itself. Thus, the triad of impairments in autism—in communication, imagination, and socialization—is explained by the failure of a single cognitive mechanism. In everyday life, even very able autistic individuals find it hard to keep in mind simultaneously a reality and the fact that someone else may hold a misconception of that reality.

The automatic ability of normal people to judge mental states enables us to be, in a sense, mind readers. With sufficient experience we can form and use a theory of mind that allows us to speculate about psychological motives for our behavior and to manipulate other people's opinions, beliefs, and attitudes. Autistic individuals lack the automatic ability to represent beliefs, and therefore they also lack a theory of mind. They cannot understand how behavior is caused by mental states or how beliefs and attitudes can be manipulated. Hence, they find it difficult to understand deception. The psychological undercurrents of real life as well as of literature—in short, all that gives spice to social relations—for them remain a closed book. "People talk to each other with their eyes," said one observant autistic youth. "What is it that they are saying?"

Lacking a mechanism for a theory of mind, autistic children develop quite differently from normal ones. Most children acquire more and more sophisticated social and communicative skills as they develop other cognitive abilities. For example, children learn to be aware that there are faked and genuine expressions of feeling. Similarly, they become adept at that essential aspect of human communication, reading between the lines. They learn how to produce and understand humor and irony. In sum, our ability to engage in imaginative ideas, to interpret feelings, and to understand intentions beyond the literal content of speech are all accomplishments that depend ultimately on an innate cognitive mechanism. Autistic children find it difficult or impossible to achieve any of these things. We believe this is because the mechanism is faulty.

This cognitive explanation of autism is specific. As a result, it enables us to distinguish the types of situations in which the autistic person will and will not have problems. It does not preclude the existence of special assets and abilities that are independent of the innate mechanism my colleagues and I see as defective. Thus it is that autistic individuals can achieve social skills that do not involve an exchange between two minds. They can learn many useful social routines, even to the extent of sometimes camouflaging their problems. The cognitive deficit we hypothesize is also specific enough not to preclude high achievement by autistic people in such diverse activities as musical performance, artistic drawing, mathematics, and memorization of facts.

It remains to be seen how best to explain the coexistence of excellent and abysmal performance by autistic people on abilities that are normally expected to go together. It is still uncertain whether there may be additional damage in emotions that prevents some autistic children from being interested in social stimuli. We have as yet little idea what to make of the single-minded, often obsessive, pursuit of certain activities. With the autistic person, it is as if a powerful integrating force—the effort to seek meaning—were missing.

The old image of the child in the glass shell is misleading in more ways than one. It is incorrect to think that inside the glass shell is a normal individual waiting to emerge, nor is it true that autism is a disorder of childhood only. The motion picture *Rain Man* came at the right time to suggest a new image to a receptive public. Here we see Raymond, a middle-aged man who is unworldly, egocentric in the extreme, and all too amenable to manipulation by others. He is incapable of understanding his brother's double-dealing pursuits, transparently obvious though they are to the cinema audience. Through various experiences it becomes possible for the brother to learn from Raymond and to forge an emotional bond with him. This is not a far-fetched story. We can learn a great deal about ourselves through the phenomenon of autism.

Yet the illness should not be romanticized. We must see autism as a devastating handicap without a cure. The autistic child has a mind that is unlikely to develop self-consciousness. But we can now begin to identify the particular types of social behavior and emotional responsiveness of which autistic individuals are capable. Autistic people can learn to express their needs and to anticipate the behavior of others when it is regulated by external, observable factors rather than by mental states. They can form emotional attachments to others. They often strive to please and earnestly wish to be instructed in the rules of person-to-person contact. There is no doubt that within the stark limitations a degree of satisfying sociability can be achieved.

Autistic aloneness does not have to mean loneliness. The chilling aloofness experienced by many parents is not a permanent feature of their growing autistic child. In fact, it often gives way to a preference for company. Just as it is possible to engineer the environment toward a blind person's needs or toward people with other special needs, so the environment can be adapted to an autistic person's needs.

On the other hand, one must be realistic about the degree of adaptation that can be made by the limited person. We can hope for some measure of compensation and a modest ability to cope with adversity. We cannot expect autistic individuals to grow out of the unreflecting mind they did not choose to be born with. Autistic people in turn can look for us to be more sympathetic to their plight as we better understand how their minds are different from our own.

—June 1993

UNDERSTANDING PARKINSON'S DISEASE

The smoking gun is still missing, but growing evidence suggests highly reactive substances called free radicals are central players in this common neurological disorder

Moussa B. H. Youdim and Peter Riederer

One of the more emotional moments of the 1996 summer Olympics in Atlanta occurred at the opening ceremonies, even before the games started. Muhammad Ali—the former world heavyweight boxing champion and a 1960 Olympic gold medal winner—took the torch that was relayed to him and, with trembling hands, determinedly lit the Olympic flame. His obvious effort reminded the world of the toll Parkinson's disease and related disorders can take on the human nervous system. Ali, who in his championship days had prided himself on his ability to "float like a butterfly, sting like a bee," now had to fight to control his body and steady his feet.

Ali's condition also highlighted the urgent need for better treatments. We cannot claim that a cure is around the corner, but we can offer a glimpse into the considerable progress investigators have made in understanding Parkinson's disease, which afflicts more than half a million people in the U.S. alone. Although still incomplete, this research has recently begun suggesting ideas not only for easing symptoms but, more important, for stopping the underlying disease process.

Parkinson's disease progressively destroys a part of the brain critical to coordinated motion. It has been recognized since at least 1817, when James Parkinson, a British physician, described its characteristic symptoms in "An Essay on the Shaking Palsy." Early on, affected individuals are likely to display a rhythmic tremor in a hand or foot, particularly when the limb is at rest. (Such trembling has helped convince many observers that Pope John Paul II has the disorder.) As time goes by, patients may become slower and stiffer. They may also have difficulty initiating movements (especially rising from a sitting position), may lose their balance and coordination, and may freeze unpredictably, as their already tightened muscles halt altogether.

Nonmotor symptoms can appear as well. These may include excessive sweating or other disturbances of the involuntary nervous system and such psychological problems as depression or, in late stages, dementia. Most of the problems, motor or otherwise, are subtle at first and worsen over time, often becoming disabling after five to 15 years. Patients typically show their first symptoms after age 60.

The motor disturbances have long been known to stem primarily from destruction of certain nerve cells that reside in the brain stem and communicate with a region underlying the cortex. More specifically, the affected neurons are the darkly pigmented ones that lie in the brain stem's substantia nigra ("black substance") and extend projections into a higher domain called the striatum (for its stripes).

As Arvid Carlsson of Gothenburg University reported in 1959, the injured neurons normally help to control motion by releasing a chemical messenger—the neurotransmitter dopamine—into the striatum. Striatal cells, in turn, relay dopamine's message through higher motion-controlling centers of the brain to the cortex, which then uses the information as a guide for determining how the muscles should finally behave. But as the dopamine-producing neurons die, the resulting decline in dopamine signaling disrupts the smooth functioning of the overall motor network and compromises the person's activity. Nonmotor symptoms apparently result mainly from the elimination of other kinds of neurons elsewhere in the brain. What remains unknown, however, is how the various neurons that are lost usually become injured.

Because damage to the substantia nigra accounts for most symptoms, investigators have concentrated on that area. Some 4 percent of our original complement of dopamine-producing neurons disappears during each decade of adulthood, as part of normal aging. But Parkinson's disease is not a normal feature of aging. A pathological process amplifies the usual cell death, giving rise to symptoms after approximately 70 percent of the neurons have been destroyed. Whether this process is commonly triggered by something in the environment, by a genetic flaw, or by some combination of the two is still unclear, although a defect on chromosome 4 has recently been implicated as a cause in some cases.

■ DRAWBACKS OF EXISTING THERAPIES

Research into the root causes of Parkinson's disease has been fueled in part by frustration over the shortcomings of the drugs available for treatment. Better understanding of the nature of the disease process will undoubtedly yield more effective agents.

The first therapeutics were found by chance. In 1867 scientists noticed that extracts of the deadly nightshade plant eased some symptoms, and so doctors began to prescribe the extracts. The finding was not explained until about a century later. By the mid-1900s pharmacologists had learned that the medication worked by inhibiting the activity in the striatum of acetylcholine, one of the chemical molecules that carries messages between neurons. This discovery implied that dopamine released into the striatum was normally needed, at least in part, to counteract the effects of acetylcholine. Further, in the absence of such moderation, acetylcholine overexcited striatal neurons that projected to higher motor regions of the brain.

Although the acetylcholine inhibitors helped somewhat, they did not eliminate most symptoms of Parkinson's disease; moreover, their potential side effects included such disabling problems as blurred vision and memory impairment. Hence, physicians were delighted when, in the 1960s, the more effective drug levodopa, or L-dopa, proved valuable. This agent, which is still a mainstay of therapy, became available thanks largely to the research efforts of Walter Birkmayer of the Geriatric Hospital Lainz-Vienna, Oleh Hornykiewicz of the University of Vienna, Theodore L. Sourkes and Andre Barbeau of McGill University, and George Cotzias of the Rockefeller University.

These and other workers developed L-dopa specifically to compensate for the decline of dopamine in the brain of Parkinson's patients. They knew that dopamine-producing neurons manufacture the neurotransmitter by converting the amino acid tyrosine to L-dopa and then converting L-dopa into dopamine. Dopamine itself cannot be used as a drug, because it does not cross the blood-brain barrier—the network of specialized blood vessels that strictly controls which substances will be allowed into the central nervous system. But L-dopa crosses the barrier readily. It is then converted to dopamine by dopamine-making neurons that survive in the substantia nigra and by nonneuronal cells, called astrocytes and microglia, in the striatum.

When L-dopa was introduced, it was hailed for its ability to control symptoms. But over time physicians realized it was far from a cure-all. After about four years, most patients experience a wearing-off phenomenon: they gradually lose sensitivity to the compound, which works for shorter and shorter increments. Also, side effects increasingly plague many people—among them, psychological disturbances and a disabling "on-off" phenomenon, in which episodes of immobility, or freezing, alternate unpredictably with episodes of normal or involuntary movements. Longer-acting preparations that more closely mimic dopamine release from neurons are now available, and they minimize some of these effects.

As scientists came to understand that L-dopa was not going to be a panacea, they began searching for additional therapies. By 1974 that quest had led Donald B. Calne and his co-workers at the National Institutes of Health to begin treating patients with drugs that mimic the actions of dopamine (termed dopamine agonists). These agents can avoid some of the fluctuations in motor control that accompany extended use of L-dopa, but they are more expensive and can produce unwanted effects of their own, including confusion, dizziness on standing, and involuntary motion.

In 1975 our own work resulted in the introduction of selegiline (also called deprenyl) for treatment of Parkinson's disease. This substance, invented by a Hungarian scientist, had failed as a therapy for depression and was almost forgotten. But it can block the breakdown of dopamine, thus preserving its availability in the striatum. Dopamine can be degraded by the neurons that make it as well as by astrocytes and microglia that reside near the site of its release. Selegiline inhibits monoamine oxidase B, the enzyme that breaks down dopamine in the astrocytes and microglia.

Selegiline has some very appealing properties, although it, too, falls short of ideal. For example, it augments the effects of L-dopa and allows the dose of that drug to be reduced. It also sidesteps the dangers of related drugs that can block dopamine degradation. Such agents proved disastrous as therapies for depression, because they caused potentially lethal disturbances in patients who ate certain foods, such as cheese. In fact, we began exploring selegiline as a treatment for Parkinson's disease partly because studies in animals had implied it would avoid this so-called cheese effect.

Tantalizingly, some of our early findings suggested that selegiline could protect people afflicted with Parkinson's disease from losing their remaining dopamine-producing neurons. A massive study carried out several years ago in the U.S. (known as DATATOP) was unable to confirm or deny this effect, but animal research continues to be highly supportive. Whether or not selegiline itself turns out to be protective, exploration of that possibility has already produced at least two important benefits. It has led to the development of new kinds of enzyme inhibitors as potential treatments not only for Parkinson's disease but also for Alzheimer's disease and depression. And the work has altered the aims of many who study Parkinson's disease, causing them to seek new therapies aimed at treating the underlying causes instead of at merely increasing the level or activity of dopamine in the striatum (approaches that relieve symptoms but do not prevent neurons from degenerating).

■ KEY ROLE FOR FREE RADICALS

Of course, the best way to preserve neurons is to halt one or more key steps in the sequence of events that culminates in their destruction—if those events can be discerned. In the case of Parkinson's disease, the collected evidence strongly implies (though does not yet prove) that the neurons that die are, to a great extent, doomed by the excessive accumulation of highly reactive molecules known as oxygen free radicals. Free radicals are destructive because they lack an electron. This state makes them prone to snatching electrons from other molecules, a process known as oxidation. Oxidation is what rusts metal and spoils butter. In the body the radicals are akin to biological bullets, in that they can injure whatever they hit—be it fatty cell membranes, genetic material, or critical proteins. Equally disturbing, by taking electrons from other molecules, one free radical often creates many others, thus amplifying the destruction.

The notion that oxidation could help account for Parkinson's disease was first put forward in the early 1970s by Gerald Cohen and the late Richard E. Heikkila of the Mount Sinai School of Medicine. Studies by others had shown that a synthetic toxin sometimes used in scientific experiments could cause parkinsonian symptoms in animals and that it worked by inducing the death of dopamine-producing neurons in the substantia nigra. Cohen and Heikkila discovered that the drug poisoned the neurons by inducing formation of at least two types of free radicals.

Some of the most direct proof that free radicals are involved in Parkinson's disease comes from examination of the brains of patients who died from the disorder. We and others have looked for "fingerprints" of free radical activity in the substantia nigra, measuring the levels of specific chemical changes the radicals are known to effect in cellular components. Many of these markers are highly altered in the brains of Parkinson's patients. For instance, we found a significant increase in the levels of compounds that form when fatty components of cell membranes are oxidized.

Circumstantial evidence is abundant as well. The part of the substantia nigra that deteriorates in Parkinson's patients contains above-normal levels of substances that promote free radical formation. (A notable example, which we have studied intensively, is iron.) At the same time, the brain tissue contains unusually low levels of antioxidants, molecules involved in neutralizing free radicals or preventing their formation.

Researchers have also seen a decline in the activity of an enzyme known as complex I in the mitochondria of the affected neurons. Mitochondria are the power plants of cells, and complex I is part of the machinery by which mitochondria generate the energy required by cells. Cells use the energy for many purposes, including ejecting calcium and other ions that can facilitate oxidative reactions. When complex I is faulty, energy production drops, free radical levels rise, and the levels of some antioxidants fall—all of which can combine to increase oxidation and exacerbate any other cellular malfunctions caused by an energy shortage.

■ EARLY CLUES FROM ADDICTS

What sequence of events might account for oxidative damage and related changes in the brains of people who suffer from Parkinson's disease? Several ideas have been proposed. One of the earliest grew out of research following up on what has been called "The Case of the Frozen Addicts."

In 1982 J. William Langston, a neurologist at Stanford University, was astonished to encounter several heroin addicts who had suddenly become almost completely immobile after taking the drug. It was as if they had developed severe Parkinson's disease overnight. While he was exploring how the heroin might have produced this effect, a toxicologist pointed him to an earlier, obscure report on a similar case in Bethesda, Maryland. In that instance, a medical student who was also a drug abuser had become paralyzed by a homemade batch of meperidine (Demerol) that was found, by Irwin J. Kopin and Sanford P. Markey of the NIH, to contain an impurity called MPTP. This preparation had destroyed dopamine-making cells of his substantia nigra. Langston, who learned that the drug taken by his patients also contained MPTP, deduced that the impurity accounted for the parkinsonism of the addicts.

His hunch proved correct and raised the possibility that a more common substance related to MPTP was the triggering cause in classical cases of Parkinson's disease. Since then, exploration of how MPTP damages dopamine-rich neurons has expanded understanding of the disease process in general and has uncovered at least one pathway by which a toxin could cause the disease.

Scientists now know that MPTP would be harmless if it were not altered by the body. It becomes dangerous after passing into the brain and being taken up by astrocytes and microglia. These cells feed the drug into their mitochondria, where it is converted (by monoamine oxidase B) to a more reactive molecule and then released to do mischief in dopamine-making neurons of the substantia nigra. Part of this understanding comes from study in monkeys of selegiline, the monoamine oxidase B inhibitor. By preventing MPTP from being altered, the drug protects the animals from parkinsonism.

In the absence of a protective agent, altered MPTP will enter nigral neurons, pass into their mitochondria, and inhibit the complex I enzyme. This action will result, as noted earlier, in an energy deficit, an increase in free radical production, and a decrease in antioxidant activity—and, in turn, in oxidative damage of the neurons.

In theory, then, an MPTP-like chemical made naturally by some people or taken up from the environment could cause Parkinson's disease through a similar process. Many workers have sought such chemicals with little success. Most recently, for instance, brain chemicals known as beta carbolines have attracted much attention as candidate neurotoxins, but their levels in the brains of Parkinson's patients appear to be too low to account for the disease. Given that years of study have not yet linked any known toxin to the standard form of Parkinson's disease, other theories may more accurately describe the events that result in excessive oxidation in patients with this disorder.

■ ARE IMMUNE CELLS OVERACTIVE?

Another hypothesis that makes a great deal of sense places microglia—the brain's immune cells—high up in the destructive pathway. This concept derives in part from the discovery, by Patrick L. McGeer of the University of British Columbia and our own groups, that the substantia nigra of Parkinson's patients often contains unusually active microglia. As a rule, the brain blocks microglia from becoming too active, because in their most stimulated state, microglia produce free radicals and behave in other ways that can be quite harmful to neurons. But if something, perhaps an abnormal elevation of certain cytokines (chemical messengers of the immune system), overcame that restraint in the substantia nigra, neurons there could well be hurt.

Studies of dopamine-making neurons conducted by a number of laboratories have recently converged with research on microglia to suggest various ways that ac-

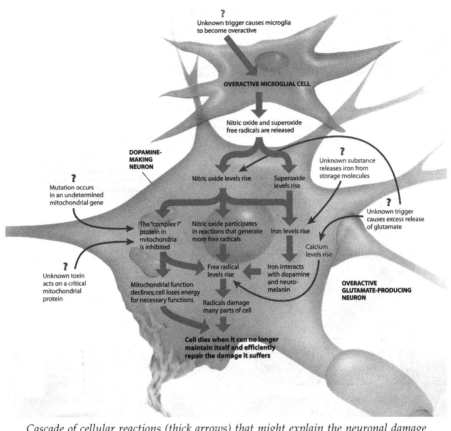

Cascade of cellular reactions (thick arrows) that might explain the neuronal damage seen in Parkinson's disease begins when some unknown signal (top) causes immune cells of the brain (microglia) to become overactive. Other as yet unidentified triggers (question marks), such as ones that overstimulate glutamate release (far right), could well initiate many of the same reactions (thin arrows). It is conceivable that Parkinson's disease sometimes results from one sequence depicted here but at other times from a combination of processes. (Alfred T. Kamajian; Laurie Grace)

tivated microglia in the substantia nigra could lead to oxidative damage in neurons of the region. Most of these ways involve production of the free radical nitric oxide.

For example, overactive microglia are known to produce nitric oxide, which can escape from the cells, enter nearby neurons, and participate in reactions that generate other radicals; these various radicals can then disrupt internal structures. Further, nitric oxide itself is able to inhibit the complex I enzyme in mitochondria; it can thus give rise to the same oxidative injury that an MPTP-like toxin could produce.

If these actions of nitric oxide were not devastating enough, we have found that both nitric oxide and another free radical (superoxide) emitted by overactive microglia can free iron from storehouses in the brain—thereby triggering additional oxidative cascades. We have also demonstrated that iron, regardless of its source, can react with dopamine and its derivatives in at least two ways that can further increase free radical levels in dopamine-synthesizing cells.

In one set of reactions, iron helps dopamine to oxidize itself. Oxidation of dopamine converts the molecule into a new substance that nigral cells use to construct their dark pigment, neuromelanin. When iron levels are low, neuromelanin serves as an antioxidant. But it becomes an oxidant itself and contributes to the formation of free radicals when it is bound by transition metals, especially iron. In support of the possibility that the interaction of iron and neuromelanin contributes to Parkinson's disease, we and our colleagues have shown that the pigment is highly decorated with iron in brains of patients who died from the disease; in contrast, the pigment lacks iron in brains of similar individuals who died from other causes.

In the other set of dopamine-related reactions, iron disrupts the normal sequence by which the neurotransmitter is broken down to inert chemicals. Neurons and microglia usually convert dopamine to an inactive substance and hydrogen peroxide, the latter of which becomes water. When iron is abundant, though, the hydrogen peroxide is instead broken down into molecular oxygen and a free radical. Dopamine's ability to promote free radical synthesis may help explain why dopamine-making neurons are particularly susceptible to dying from oxidation. This ability has also contributed to suspicion that L-dopa, which increases dopamine levels and eases symptoms, may, ironically, damage nigral neurons. Scientists are hotly debating this topic, although we suspect the concern is overblown.

In brief, then, overactive microglia could engender the oxidative death of dopamine-producing neurons in the substantia nigra by producing nitric oxide, thereby triggering several destructive sequences of reactions. And iron released by the nitric oxide or other free radicals in the region could exacerbate the destruction. As we have noted, brain cells do possess molecules capable of neutralizing free radicals. They also contain enzymes that can repair oxidative damage. But the protective systems are less extensive than those elsewhere in the body and, in any case, are apparently ill equipped to keep up with an abnormally large onslaught of oxidants. Consequently, if the processes we have described were set off in the substantia nigra, one would expect to see ever more neurons fade from the region over time, until finally the symptoms of Parkinson's disease appeared and worsened.

Actually, any trigger able to induce an increase in nitric oxide production or iron release or a decrease in complex I activity in the substantia nigra would promote Parkinson's disease. Indeed, a theory as plausible as the microglia hypothesis holds that excessive release of the neurotransmitter glutamate by neurons feeding into the striatum and substantia nigra could stimulate nitric oxide production and iron release. Excessive glutamate activity could thus set off the same destructive cascade hypothetically induced by hyperactive microglia. Overactive glutamate release has been implicated in other brain disorders, such as stroke. No one yet knows whether glutamate-producing neurons are overactive in Parkinson's disease, but circumstantial evidence implies they are.

Other questions remain as well. Researchers are still in the dark as to whether Parkinson's disease can arise by different pathways in different individuals. Just as the engine of a car can fail through any number of routes, a variety of processes could presumably lead to oxidative or other damage to neurons of the substantia nigra. We also have few clues to the initial causes of Parkinson's disease—such as triggers that might, say, elevate cytokine levels or cause glutamate-emitting cells to be hyperactive. In spite of the holes, ongoing research has suggested intriguing ideas for new therapies aimed at blocking oxidation or protecting neurons in other ways.

■ THERAPEUTIC OPTIONS

If the scenarios we have discussed do occur alone or together, it seems reasonable to expect that agents able to quiet microglia or inhibit glutamate release in the substantia nigra or striatum would protect neurons in at least some patients. The challenge is finding compounds that are able to cross the blood-brain barrier and produce the desired effects without, at the same time, disturbing other neurons and causing severe side effects. One of us (Riederer) and his colleague Johannes Kornhuber of the University of Würzburg have recently demonstrated that amantadine, a long-standing anti-Parkinson's drug whose mechanism of action was not known, can block the effects of glutamate. This result suggests that the compound might have protective merit. Another glutamate blocker—dextromethorphan—is in clinical trials at the NIH.

Drugs could also be protective if they halted other events set in motion by the initial triggers of destruction. Iron chelators (which segregate iron and thus block many oxidative reactions), inhibitors of nitric oxide formation, and antioxidants are all being considered. Such agents have been shown to protect dopamine-producing neurons of the substantia nigra from oxidative death in animals. On the other hand, the same human DATATOP trial that cast doubt on selegiline's protective effects found that vitamin E, an antioxidant, was ineffective. But vitamin E may have failed because very little of it crosses the blood-brain barrier or because the doses tested were too low. Antioxidants that can reach the brain deserve study; at least one such compound is in clinical trials at the NIH.

Regardless of the cause of the neuronal destruction, drugs that were able to promote regeneration of lost neurons would probably be helpful as well. Studies of animals suggest that such substances could, indeed, be effective in the human brain. Researchers at several American facilities are now testing putting a molecule called glial-derived neurotrophic factor (GDNF) directly into the brain of patients. Efforts are also under way to find smaller molecules that can be delivered more conveniently (via pill or injection) yet would still activate neuronal growth factors and neuronal growth in the brain. One agent, Rasagiline, has shown promise in animal trials and is now being tested in humans. Some studies imply that the nicotine in tobacco might have a protective effect, and nicotinelike drugs are being studied in the laboratory as potential therapies. Patients, however, would be foolish to take up smoking to try to slow disease progression. Data on the value of smoking to retard the death of dopamine neurons are equivocal, and the risks of smoking undoubtedly far outweigh any hypothetical benefit.

As work on protecting neurons advances, so does research into compensating for their decline. One approach being perfected is the implantation of dopamine-producing cells. Some patients have been helped. But the results are variable, and cells available for transplantation are in short supply. Further, the same processes that destroyed the original brain cells may well destroy the implants. Other approaches include surgically destroying parts of the brain that function abnormally when dopamine is lost. This surgery was once unsafe but is now being done more successfully.

The true aim of therapy for Parkinson's disease must ultimately be to identify the disease process long before symptoms arise, so that therapy can be given in time to forestall the brain destruction that underlies patients' discomfort and disability. No one can say when early detection and neural protection will become a reality, but we would not be surprised to see great strides made on both fronts within a few years. In any case, researchers cannot rest easy until those dual objectives are met.

—January 1997

Amyloid Protein and Alzheimer's Disease

When this protein fragment accumulates excessively in the brain, Alzheimer's disease may be the result. Understanding how that fragment forms could be the key to a treatment

Dennis J. Selkoe

Imagine surmounting life's great and small hurdles only to face at the end the relentless and devastating loss of one's most human qualities: reasoning, abstraction, language, and memory. Such a fate now awaits millions of individuals in races and ethnic groups worldwide. The dramatic rise in life expectancy during this century, primarily through the cure of infectious diseases, has enabled many of us to reach an age at which degenerative diseases of the brain—particularly Alzheimer's disease—become common.

Most of us were raised on the notion that a grandparent who became confused and forgetful had "hardening of the arteries." Dementia, the failure of cognitive ability, was widely assumed to be a natural accompaniment of old age. But neuropathological studies during the past two decades have shown that the brain lesions first described by the Bavarian psychiatrist Alois Alzheimer in 1907—senile plaques and neurofibrillary tangles—are the most common basis for late-life dementia in many developed countries.

The loss of memory, judgment, and emotional stability that Alzheimer's disease inflicts on its victims occurs gradually and inexorably, usually leading to death in a severely debilitated, immobile state between four and 12 years after onset. The cost to American society for diagnosing and managing Alzheimer's disease, primarily for custodial care, is currently estimated at more than $80 billion annually. No treatment that retards the progression of the disease is known.

How should scientists interested in deciphering and ultimately blocking this complex disorder begin their attack? The answer appears increasingly to lie in understanding the genesis of its hallmark: the so-called senile plaques that occur in huge numbers within the cerebral cortex, the hippocampus, the amygdala, and other brain areas essential for cognitive function. Recent discoveries indicate that such work is indeed providing powerful clues about the earliest events in the mechanism of the disease.

Peering through the microscope at the brain of his first patient, Alzheimer wrote prophetically: "Scattered through the entire cortex, especially in the upper layers, one found miliary foci that were caused by the deposition of a peculiar substance in the cerebral cortex." Evidence emerging from many laboratories during the past seven years indicates that this "peculiar substance" is a protein fragment, approximately 40 amino acids long, referred to as the amyloid beta-protein. It arises from cleavage by enzymes of a much larger precursor protein encoded by a gene on human chromosome 21.

The study of the amyloid beta-protein has helped to clarify the genetic basis of Alzheimer's disease. It has long been known that at least some cases are caused by genetic abnormalities: in some families, roughly one half of each generation acquires this familial type of Alzheimer's disease. People with Down's syndrome, who are born with three copies of chromosome 21 instead of the normal two, almost always acquire the brain lesions typical of Alzheimer's disease prematurely, by their forties or fifties. The behavior and mental abilities of many Down's syndrome patients seem to decline further at about the same time.

Discoveries about the genetic mechanisms regulating the accumulation of amyloid beta-protein are helping us understand why past attempts to treat Alzheimer's disease have been so fruitless—and they are steering us toward new treatments that may be significantly more successful.

The story of the beta-protein in Alzheimer's disease begins with the neuropathologist. Even before the time of Alzheimer, pathologists knew that the human cerebral cortex sometimes contained variable numbers of spherical plaques. These plaques consisted of altered axons and dendrites (the long, tapering ends of neurons, collectively called neurites) surrounding an extracellular mass of thin filaments. Under the microscope, these filaments resembled similar extracellular deposits that accumulated in other organs in a variety of unrelated diseases.

In 1853 the great German pathologist Rudolf Virchow called such deposits "amyloid," an unfortunate term because it implied that the deposits were made of a starchlike substance. Chemical studies have shown that the principal constituents of the amyloid filaments are actually proteins and that the identity of the proteins differs among the various diseases marked by the deposition of amyloid. The common thread among these disparate diseases, or amyloidoses, is that they are characterized by innumerable extracellular deposits of normal or mutated protein fragments. Moreover, the protein subunits are always folded in a particular three-dimensional pattern called a beta-pleated sheet.

The senile plaque is a complex, slowly evolving structure, and the time required to generate fully formed, "mature" plaques may be years or even decades. In addition to the central core of amyloid beta-protein and the surrounding abnormal neurites, mature plaques contain two types of altered glial cells. (Glial cells normally associate with neurons and perform many supportive and protective functions.) In the center of the mature plaques, one usually observes microglial cells, which are scavengers capable of responding to inflammation or the destruction of nervous system tissue in many brain disorders. Around the outside of the plaque are so-called reactive astrocytes, glial cells that are often found in injured brain areas.

Along with the senile plaques, brain tissue affected by Alzheimer's disease is characterized by variable numbers of neurofibrillary tangles: dense bundles of abnormal fibers in the cytoplasm of certain neurons. These fibers, or paired helical filaments, are not made of the amyloid beta-protein. Instead they appear to be composed of a modified form of a normally occurring neuronal protein called tau. Like the amyloid plaques, the neurofibrillary tangles are not specific for Alzheimer's disease. The tangles occur in a dozen or more chronic diseases of the human brain.

Most of us who live into our late seventies will develop at least a few senile plaques and neurofibrillary tangles, particularly in the hippocampus and other brain regions important for memory. For the most part, the distinction between normal brain aging and Alzheimer's disease is quantitative rather than qualitative. Usually patients with progressive dementia of the Alzheimer type have moderately or markedly more mature neuritic plaques and neurofibrillary tangles than age-matched nondemented people do. Elucidation of the genesis of the plaques and tangles in Alzheimer's disease should therefore tell us a lot about the highly similar lesions that underlie in part the more subtle changes in memory and cognition affecting some otherwise healthy septuagenarians.

Because large numbers of amyloid-bearing neuritic plaques in brain regions critical for intellectual function are an invariant feature of Alzheimer's disease, we need to understand the nature of the amyloid protein. In 1984 George G. Glenner and Caine W. Wong of the University of California at San Diego first isolated amyloid from blood vessels in the meninges (connective tissue surrounding the brain) of Alzheimer patients. When they solubilized the isolated amyloid, they found it was composed of a small protein, dubbed the "beta protein," that had a novel amino acid sequence.

Shortly thereafter, Colin L. Masters of the University of Western Australia, Konrad Beyreuther of the University of Cologne, and their colleagues isolated the amyloid cores of senile plaques. The core protein had the same size and amino acid composition as the meningovascular beta-protein, and antibodies against one reacted with the other.

But plaque cores isolated by Carmela Abraham and me at that time yielded beta-protein that had a "blocked" first amino acid and thus could not be sequenced, in contrast to the vessel-derived beta-protein that had been successfully sequenced by Glenner and Wong. This and other findings suggested to us that the mature plaque core contained beta-protein that was chemically modified as compared with that found in blood vessels. It appears that both vascular and plaque amyloid are composed of the beta-protein but that its precise chemical form in the two sites may differ slightly.

After a protein is purified and sequenced, the next step in its characterization is often the cloning of the complementary DNA molecule that embodies the genetic code for synthesizing the protein. In the case of amyloid beta-protein, this was accomplished independently by four laboratories in early 1987. One of these groups—Jin Kang, Beyreuther, Benno Müller-Hill, and their colleagues at the University of Cologne—had the good fortune of isolating a stretch of DNA that contained the entire coding sequence for the protein. This sequence demonstrated that amyloid beta-protein was just a small fragment, 40 or so amino acids, out of a 695-amino acid protein, which is now generally referred to as the beta-amyloid precursor protein (beta-APP).

The presumed structure of beta-APP included one region (from amino acid 625 to 648) that could anchor the molecule to cell membranes. To everyone's surprise, the small amyloid beta-protein fragment comprised amino acids 597 through 636—the 28 amino acids just outside the membrane-spanning domain and the first 12 amino acids within the membrane. That finding presented a conundrum that has still not been resolved: How can a segment of beta-APP that normally anchors it firmly to cell membranes appear in the extracellular space as amyloid? Or to put the question another way, how can the enzymes that snip amyloid beta-protein out of its large precursor gain access to that transmembrane region? The answer may have important implications for blocking the amyloidosis of Alzheimer's disease.

Perhaps even more exciting than knowing the amino acid structure of the amyloid precursor was the finding, made by each of the four laboratories that had cloned beta-APP, that the precursor was encoded by a gene located on chromosome 21. This discovery gave rise to a kind of global "aha!" experience in the Alzheimer field. We could suddenly fathom why people with Down's syndrome, who are born with an extra copy of that chromosome, routinely developed beta-amyloid deposits at a relatively early age. At about the same time a collaborative group led by Peter St. George-Hyslop, Rudolph Tanzi, and James F. Gussella of the Massachusetts General Hospital found that at least one form of familial Alzheimer's disease (FAD) appeared to be caused by a genetic defect that was also located somewhere on chromosome 21.

At first it seemed that the gene for beta-APP might itself be the one responsible for early-onset FAD. Yet as so often happens in science, the situation is more complex. Several studies of FAD patients in late 1987 and early 1988 failed to reveal either defects or duplications in the protein-coding region of the beta-APP gene; also, a long stretch of DNA appeared to separate the beta-APP gene from the approximate site of the early-onset FAD defect on chromosome 21. Moreover, Gerard D. Schellenberg and his collaborators at the University of Washington and Margaret A. Pericak-Vance and her colleagues at Duke University Medical School were unable to demonstrate linkage of Alzheimer's disease to markers on chromosome 21 in several different families, including some with late-onset (older than 65 years) forms of the disease. These and other studies have led to the conclusion that FAD is genetically heterogeneous—it can apparently be caused by many different genetic defects on different chromosomes.

This realization should perhaps come as no surprise. Alzheimer's disease is very common, occurs in many different ethnic groups, and is closely similar to the normal process of brain aging. It is reasonable that the syndrome we call Alzheimer's disease could arise in somewhat different forms from many distinct genetic alterations. All these alterations, however, appear to act through a critical common mechanism involving the increased deposition of amyloid beta-protein.

A new and potentially exciting development in the search for the genes that cause FAD was reported recently by Alison Goate, John Hardy, and their collaborators at St. Mary's Hospital in London. They found two families in which everyone who had Alzheimer's disease also had a particular DNA pattern in the beta-APP gene. When the investigators sequenced the DNA encoding the amyloid beta-protein region of beta-APP, they found a mutation that resulted in a switch of amino acid 642 (out of 695) from a valine to an isoleucine. Because no mutation in the beta-APP gene had previously been found in either normal or FAD subjects, that change is probably not an irrelevant chance event but rather the cause of Alzheimer's disease in these two families. In recent months at least six other families prone to FAD that have a DNA mutation at position 642 have been found.

Two major conclusions emerge. First, one specific molecular cause of Alzheimer's disease has been identified. Second, it is now clear that beta-amyloid deposition can arise directly from a mutation in beta-APP, without any other preexisting cellular or molecular defect. An idea that some of us have long held—that beta-amyloid abnormalities can truly initiate some forms of Alzheimer's disease—has moved from speculation to reality.

Some cases of Alzheimer's disease appear to occur sporadically, that is, without any known familial predisposition. It is difficult, however, to reach a firm conclusion in this regard about a late-onset illness, because family members bearing a faulty gene may have died of other causes before the symptoms appeared. Environmental factors are likely to influence the onset of Alzheimer's disease. One piece of evidence supporting this opinion is the observation that identical twins may manifest Alzheimer symptoms at considerably different ages. Unfortunately, the search for environmental factors that can trigger the disease has been resoundingly unsuccessful to date, although debate still swirls around the unsettled role of aluminum as a contributing factor. One possible environmental trigger noted in a small minority of patients is a history of major head trauma, although how trauma could accelerate beta-amyloid deposition is unclear.

While investigations of the genetic defects underlying familial Alzheimer's disease have gone forward, steady progress in characterizing beta-APP itself and the role of the beta-protein in the disease has also continued. On the molecular biological front, several laboratories discovered alternative DNA sequences encoding beta-APP—sequences that contain either one or two additional coding segments at position 289 of the original 695-amino acid form. One of these two inserts encodes a stretch of amino acids that has the ability to bind to and inhibit proteases, the enzymes that cut proteins into smaller fragments. Because of that discovery, we can guess for the first time about one normal function of beta-APP: it may be an inhibitory molecule that regulates the activity of proteases.

On the protein chemistry front, Marcia Podlisny and I at Harvard, in collaboration with Tilman Oltersdorf and Lawrence C. Fritz of Athena Neurosciences in South San Francisco, identified and characterized beta-APP in brain and other human tissues and in cultured cells. In all the tissues that we examined, we detected a stable fragment of beta-APP that contained one end (the carboxyl terminus) of the molecule as well as part or all of the critical amyloid beta-protein region. Several laboratories subsequently showed that the other part of beta-APP, the one with the amino terminus, is shed into the fluid outside of cells, including cerebrospinal fluid and plasma. In 1990 Fred Esch and his co-workers at Athena Neurosciences showed that this normally occurring fragmentation of beta-APP resulted from a cleavage of the full-length precursor at amino acid 16 within the amyloid beta-protein region.

Their finding tells us that the as yet unidentified protease that normally cuts at this site prevents the formation of intact amyloid beta-protein. The beta-amyloid deposition that occurs during aging and in Alzheimer's disease must therefore involve an alternative proteolytic pathway—one that cleaves beta-APP at the beginning and end of the amyloid beta-protein region. Researchers are now avidly seeking those alternative proteases, because drugs designed to inhibit them should decrease or prevent the deposition of amyloid beta-protein.

I have placed great emphasis on understanding the normal and abnormal processing of beta-APP. Yet what is the evidence that amyloid beta-protein deposition precedes the pathology of Alzheimer's disease rather than follows it? Since the beginning of the century, neuropathologists have argued about whether the amyloid in the core of the senile plaque was produced by the neurites in the plaque periphery as they degenerated or whether its appearance preceded and caused neuritic changes. Some researchers wondered whether the amyloid might have come from healthy neurons or from glial cells or even from nearby blood vessels.

In 1988 and 1989 several investigators noticed that amorphous, nonfilamentous deposits of amyloid beta-protein occur in Alzheimer brain tissue and that such diffuse, or "preamyloid," plaques are actually much more abundant than the classic neuritic plaques. Using antibodies against amyloid beta-protein as highly sensitive probes, researchers have detected such diffuse plaques not only in the brain areas implicated in the symptoms of the disease, such as the cerebral cortex, but also in other areas, such as the thalamus and cerebellum. Significantly, most diffuse beta-protein deposits contain few or no degenerating neurites or reactive glial cells. Electron microscopic examinations by Haruyasu Yamaguchi of Gunma University in Japan have revealed that much of the tissue within the diffuse plaque is indistinguishable from surrounding normal brain tissue.

Several laboratories have now also reported that some patients with Down's syndrome who die in their teens or twenties seem to have many diffuse plaques in the absence of mature neuritic plaques, neurofibrillary tangles, or other signs of cellular pathology. Because virtually all such patients would ultimately have exhibited these full-blown lesions, one may conclude that diffuse amyloid beta-protein deposits precede Alzheimer-type neuropathology in Down's syndrome and, by implication, in Alzheimer's disease itself. It is likely that only a minority of the diffuse beta-protein deposits gradually progresses to involve the surrounding neurites and glia. For unknown reasons, this maturation seems to occur much more commonly in the symptom-producing cerebral cortex than in, for example, the symptom-free cerebellum.

The observation of diffuse plaques suggested that amyloid beta-protein deposition preceded the alteration of neurons and other brain cells. Such data, as well as the occurrence of amyloid beta-protein in the walls of meningeal blood vessels outside brain tissue, led my Harvard Medical School colleagues Catharine L. Joachim and Hirshi Mori and me to search for amyloid beta-protein deposition in organs other than the brain. In early 1989 we found small deposits in and around selected blood vessels of the skin, intestine, and certain other tissues from some Alzheimer patients and aged control subjects; these deposits reacted specifically with antibodies against amyloid beta-protein.

The detection of those deposits provided the first evidence that the process underlying beta-amyloid deposition in Alzheimer's disease may not be restricted to the brain. The predilection of these extracerebral deposits to occur near blood vessels also strengthens the parallels between Alzheimer's disease and certain systemic amyloidoses that we know have a circulatory origin. Most important, the deposition of small amounts of amyloid beta-protein in peripheral blood vessels in the absence of any preceding neuronal injury—indeed in the absence of local neurons and glial cells—supports the hypothesis that the release and accumulation of amyloid beta-protein precede rather than follow neuronal degeneration in the brain.

Some of the clearest evidence that amyloid beta-protein deposition can be the primary event initiating an illness has emerged from studies of a rare genetic disorder found in two villages in the Netherlands. Patients in the affected families die in midlife from cerebral hemorrhages caused by severe amyloid deposition in innumerable blood vessels, hence the name "hereditary cerebral hemorrhage with amyloidosis of the Dutch type" (HCHWA-Dutch). Blas Frangione and his collaborators at New York University Medical Center and the University of Leiden found that the deposited protein in this disease was in fact amyloid beta-protein.

Then, in 1990, Efrat Levy and Frangione of New York University, Mark Carman of Athena Neurosciences, Sjoerd van Duinen of the University of Leiden, and their co-workers discovered a DNA mutation in the beta-APP gene that caused the substitution of the amino acid glutamine for a glutamic acid at position 22 within the amyloid beta-protein. Exactly why this unique mutation leads to such severe cerebrovascular deposition is not clear. Nevertheless, genetic analyses of affected and unaffected siblings by Christine Van Broeckhoven of the University of Antwerp and her collaborators confirm that the mutation is the disease-causing defect.

The Dutch patients also seem to show numerous beta-protein deposits in the cerebral cortex that closely resemble the preamyloid plaques of Alzheimer's disease and Down's syndrome. For unknown reasons, those deposits do not seem to affect the surrounding neurites and glia: no neurofibrillary tangles or other neuronal alterations appear, and the patients do not suffer Alzheimer-type dementia. The characterization of HCHWA-Dutch nonetheless demonstrated for the first time that a mutation in beta-APP can cause diffuse amyloid beta-protein plaques and cerebrovascular beta-amyloidosis. The work provides a compelling example of a principle of medical research: the study of a rare disease in a small patient group can provide critical insights into common pathological processes that affect the entire population.

Even if one assumes that these various observations support the early deposition of amyloid beta-protein in Alzheimer's disease, what real evidence is there that the beta-protein is biologically active? Researchers must determine whether amyloid beta-protein is itself injurious to neurons or whether it serves as a matrix to which other more active molecules bind. At the moment, only fragmentary information about the biological effects of amyloid beta-protein is available. In 1988 Janet S. Whitson and Carl W. Cotman of the University of California at Irvine, in collaboration with me, noticed that the first 28 amino acids of the amyloid beta-protein had trophic (survival-promoting) effects on cultures of rat hippocampal neurons. Subsequently, Cotman and his colleagues demonstrated that the entire beta-protein molecule had both neurite-promoting effects and indirect neurotoxic effects.

More recently, Bruce A. Yankner and his colleagues at the Children's Hospital in Boston have reported that, under the right conditions, low doses of the full-length amyloid beta-protein can enhance the survival of freshly cultured rat neurons. When those investigators raised the dose modestly and allowed the cultures to age for four days, however, they observed toxic effects on some of the neurons. They localized these biological effects to one segment of the amyloid beta-protein (amino acids 25 through 35). The amino acid sequence of this molecular region is very similar to that of a naturally occurring brain peptide called substance P. If the neurotoxic effects observed in Yankner's study are confirmed, the argument that amyloid beta-protein is directly responsible for the neuronal pathology found in neuritic plaques will be substantially strengthened.

A major obstacle to deciphering how the beta-protein is released from its large precursor and causes neurons and glial cells to change is the lack of a close, convenient animal model for Alzheimer's disease. Inexpensive laboratory rodents do not spontaneously develop cerebral amyloid deposits with age, although expensive primates do. In 1987 my colleagues and I, in collaboration with Linda Cork and Donald Price of the Johns Hopkins School of Medicine, showed that antibodies made against purified human beta-protein reacted strongly with the amyloid found in the senile plaques and blood vessels of old monkeys, dogs, and several other mammals. Additional studies in Price's laboratory and ours have yielded evidence that monkeys undergo an age-related beta-amyloidosis strikingly similar to that of humans. When drugs that block either the formation or the activity of the beta-protein are developed, it will be important to test their efficacy and safety in primates.

A more practical animal model is now being developed by injecting fragments of human DNA into fertilized mouse ova. Several laboratories have recently reported the production and breeding of these transgenic mice, which incorporate part of the human beta-APP gene into their genomes. In one example, transgenic mice developed by Barbara Cordel and her colleagues at California Biotechnology exhibit deposits of beta-protein that show some of the features of diffuse plaques. Whether such deposits will lead to alterations of the surrounding neurons and glial cells that are analogous to those of Alzheimer's disease remains to be seen. Various strains of transgenic mice expressing the mutant beta-APP molecules will be particularly important for studying the formation of beta-protein deposits and their inhibition by drugs.

How can we integrate the rapidly emerging knowledge about the nature and effects of amyloid beta-protein and its precursor into a dynamic model of how at least some forms of Alzheimer's disease work? I speculate that a group of distinct but related defects on chromosome 21, including DNA mutations, deletions, and perhaps rearrangements, can result in either structurally abnormal beta-APP proteins or deregulation of the transcription of the beta-APP gene.

In families with the latter type of defect, the DNA that is altered presumably controls how much or what form of messenger RNA is transcribed from the beta-APP gene. (Messenger RNAs are the crucial intermediary molecules in the translation of proteins from genes.) The control of gene transcription is a highly complex process, and the regulatory elements within the large beta-APP gene are not yet well understood. For example, a DNA alteration on chromosome 21 that causes some forms of FAD might interact adversely with the DNA in the regulatory regions of the beta-APP gene; alternatively, a mutation might occur directly within one of the regulatory regions. In the model I propose, the ultimate result of those DNA defects that occur outside the protein coding region of the beta-APP gene is to enhance—sometimes subtly, sometimes robustly—the amount or type of beta-APP proteins synthesized in at least some cell types. The result might be an overproduction of beta-APP somewhat reminiscent of that in Down's syndrome, which occurs because of an extra copy of the beta-APP gene.

The identity of the cell types producing the beta-APP molecules that release the amyloid beta-protein is unknown. In my view, circulating blood cells (such as platelets) and endothelial cells lining the blood vessel wall are among the most likely candidates, although neurons and glial cells are also reasonable possibilities. If these cells synthesized either excess or altered forms of beta-APP, some of those molecules might be broken down by an alternative enzymatic pathway, thereby liberating large fragments that contain the amyloid beta-protein.

Over time, I suspect, these fragments are further cleaved by proteases to release the intact amyloid beta-protein, which then accumulates in the extracellular spaces of the brain in the form of diffuse plaques. Because of local tissue factors in the cerebral cortex and other brain regions important for cognitive function, a minority of these diffuse plaques becomes increasingly filamentous and compact. The addition of so-called beta-amyloid-associated proteins—some of which have already been identified—and the activation of nearby microglia and astrocytes probably contribute to the maturation of the plaques.

At some point in this dynamic process, amyloid beta-protein or other molecules attracted to it begin to have both trophic and toxic effects on the surrounding axons and dendrites, resulting in a further evolution of some plaques to a neuritic form. In all probability, many other biochemical and structural changes accompany this phase, including a loss of synapses and a resultant decline in the cortical levels of acetylcholine and some other neurotransmitters. Some of the affected neurons would produce the masses of paired helical filaments that constitute the neurofibrillary tangles. As this complex and progressive cascade of molecular changes proceeds in areas such as the hippocampus and cerebral cortex, the patient slowly acquires symptoms of intellectual failure.

The mechanism of the disease I have outlined here is both speculative and simplified. Several observations cannot yet be placed into this scheme. The greatest challenge to students of Alzheimer's disease—and the greatest source of ongoing controversy—is the attempt to arrange the observed biochemical changes in a temporal sequence of pathogenesis. In proposing this model of Alzheimer's disease, particularly of the familial type, I have emphasized alterations in the regulation of the synthesis of beta-APP. The chromosome 21–linked form of FAD is strikingly similar to Down's syndrome, in which the elevated synthesis of beta-APP is a fact, not speculation.

Nevertheless, defects in the metabolism and degradation of beta-APP could also underlie some cases of Alzheimer's disease. For example, genetic defects could alter the enzymes that are normally responsible for attaching carbohydrate or phosphate molecules to beta-APP, or they could alter the proteases that normally break down beta-APP. In such cases, the gene making the defective enzyme could be located on any of the chromosomes, not just on chromosome 21. To date, no other chromosome has been clearly implicated as the site of a gene defect causing FAD, but Allen D. Roses and his colleagues at Duke University Medical Center have raised the possibly that in some families a gene on chromosome 19 may be involved.

In addition to searching for genetic alterations that may underlie different cases of FAD, it is important to pursue epidemiologic surveys that might uncover environmental factors that predispose individuals to Alzheimer's disease or that accelerate its course. At present, there is no compelling evidence that factors such as nutrition, educational level, occupation, or emotional state influence the occurrence or progression of the disease.

The goal of all Alzheimer's disease research is the development of an effective therapy. In 1976 three laboratories in Great Britain published evidence that neurons synthesizing the neurotransmitter acetylcholine were severely compromised in the cerebral cortex of Alzheimer patients. Physicians subsequently attempted to increase the amount of acetylcholine by administering acetylcholine-enhancing drugs, largely to no avail. Today we know that Alzheimer's disease affects many different types of neurons and neurotransmitters, which is why the illness is difficult to treat by simply replacing the neurotransmitters.

The model for Alzheimer's disease that I have described point to new therapeutic strategies aimed at one or more crucial steps in its molecular progression. First, one might block the delivery to the cerebrum and its vasculature of those beta-APP molecules that are responsible for the amyloid deposits. That approach would be most feasible if, as I suspect, those proteins arrive by way of the bloodstream. Second, one could inhibit the proteases that liberate amyloid beta-protein by cleaving beta-APP. Third, one could retard the apparent maturation of amyloid beta-protein deposits into neuritic plaques, perhaps by interfering with the formation of the amyloid filaments that seems to accompany this change. Fourth, one could interfere with the activities of the microglia, astrocytes, and other cells that contribute to the chronic inflammation around the neuritic plaques. And fifth, one might block the molecules on the surface of neurons that mediate the trophic and toxic effects of amyloid beta-protein and the proteins associated with it in the plaques.

None of these pharmacological targets will be easy to reach, but the number of investigators focusing on beta-amyloidosis and the current pace of discovery make it likely that inhibitors of one or more crucial steps in the development of the disease will emerge in the next few years. In view of the tragedy that Alzheimer's disease represents for its victims and for society, this feat cannot come a moment too soon.

—November 1991

THE NEUROBIOLOGY
OF DEPRESSION

The search for biological underpinnings of depression is intensifying. Emerging findings promise to yield better therapies for a disorder that too often proves fatal

Charles B. Nemeroff

In his 1990 memoir *Darkness Visible,* the American novelist William Styron—author of *The Confessions of Nat Turner* and *Sophie's Choice*—chillingly describes his state of mind during a period of depression:

> He [a psychiatrist] asked me if I was suicidal, and I reluctantly told him yes. I did not particularize—since there seemed no need to—did not tell him that in truth many of the artifacts of my house had become potential devices for my own destruction: the attic rafters (and an outside example or two) a means to hang myself, the garage a place to inhale carbon monoxide, the bathtub a vessel to receive the flow from my opened arteries. The kitchen knives in their drawers had but one purpose for me. Death by heart attack seemed particularly inviting, absolving me as it would of active responsibility and I had toyed with the idea of self-induced pneumonia—a long frigid, shirt-sleeved hike through the rainy woods. Nor had

I overlooked an ostensible accident, à la Randall Jarrell, by walking in front of a truck on the highway nearby. . . . Such hideous fantasies, which cause well people to shudder, are to the deeply depressed mind what lascivious daydreams are to persons of robust sexuality.

As this passage demonstrates, clinical depression is quite different from the blues everyone feels at one time or another and even from the grief of bereavement. It is more debilitating and dangerous, and the overwhelming sadness combines with a number of other symptoms. In addition to becoming preoccupied with suicide, many people are plagued by guilt and a sense of worthlessness. They often have difficulty thinking clearly, remembering, or taking pleasure in anything. They may feel anxious and sapped of energy and have trouble eating and sleeping or may, instead, want to eat and sleep excessively.

Psychologists and neurobiologists sometimes debate whether ego-damaging experiences and self-deprecating thoughts or biological processes cause depression. The mind, however, does not exist without the brain. Considerable evidence indicates that regardless of the initial triggers, the final common pathways to depression involve biochemical changes in the brain. It is these changes that ultimately

---

THE SYMPTOMS OF MAJOR DEPRESSION

The American Psychiatric Association considers people to have the syndrome of clinical depression if they show five or more of the following symptoms nearly every day during the same two-week span. The symptoms must include at least one of the first two criteria, must cause significant distress or impairment in daily functioning, and cannot stem from medication, drug abuse, a medical condition (such as thyroid abnormalities), or uncomplicated bereavement. For the formal criteria, see the association's *Diagnostic and Statistical Manual of Mental Disorders,* fourth edition.

- Depressed moods most of the day (in children and adolescents, irritability may signify a depressed mood)

- Markedly diminished interest or pleasure in all or most activities most of the day

- Large increase or decrease in appetite

- Insomnia or excessive sleeping

- Restlessness (evident by hand wringing and such) or slowness of movement

- Fatigue or loss of energy

- Feelings of worthlessness or excessive or inappropriate guilt

- Indecisiveness or diminished ability to think or concentrate

- Recurrent thoughts of death or of suicide

give rise to deep sadness and the other salient characteristics of depression. The full extent of those alterations is still being explored, but in the past few decades—and especially in the past several years—efforts to identify them have progressed rapidly.

At the moment, those of us teasing out the neurobiology of depression somewhat resemble blind searchers feeling different parts of a large, mysterious creature and trying to figure out how their deductions fit together. In fact, it may turn out that not all of our findings will intersect: biochemical abnormalities that are prominent in some depressives may differ from those predominant in others. Still, the extraordinary accumulation of discoveries is fueling optimism that the major biological determinants of depression can be understood in detail and that those insights will open the way to improved methods of diagnosing, treating, and preventing the condition.

■ PRESSING GOALS

One subgoal is to distinguish features that vary along depressed individuals. For instance, perhaps decreased activity of a specific neurotransmitter (a molecule that carries a signal between nerve cells) is central in some people, but in others, overactivity of a hormonal system is more influential (hormones circulate in the blood and can act far from the site of their secretion). A related goal is to identify simple biological markers able to indicate which profile fits a given patient; those markers could consist of, say, elevated or reduced levels of selected molecules in the blood or changes in some easily visualizable areas of the brain.

After testing a depressed patient for these markers, a psychiatrist could, in theory, prescribe a medication tailored to that individual's specific biological anomaly; much as a general practitioner can run a quick strep test for a patient complaining of a sore throat and then prescribed an appropriate antibiotic if the test is positive. Today psychiatrists have to choose antidepressant medications by intuition and trial and error, a situation that can put suicidal patents in jeopardy for weeks or months until the right compound is selected. (Often psychotherapy is needed as well, but it usually is not sufficient by itself, especially if the depression is fairly severe.)

Improving treatment is critically important. Although today's antidepressants have fewer side effects than those of old and can be extremely helpful in many cases, depression continues to exact a huge toll in suffering, lost lives, and reduced productivity.

The prevalence is surprisingly great. It is estimated, for example, that 5 to 12 percent of men and 10 to 20 percent of women in the U.S. will suffer from a major depressive episode at some time in their life. Roughly half of these individuals will become depressed more than once, and up to 10 percent (about 1.0 to 1.5 percent of Americans) will experience manic phases in addition to depressive ones, a condition known as manic-depressive illness or bipolar disorder. Mania is marked by a decreased need for sleep, rapid speech, delusions of grandeur, hyperactivity, and a propensity to engage in such potentially self-destructive activities as promiscuous sex, spending sprees, or reckless driving.

Beyond the pain and disability depression brings, it is a potential killer. As many as 15 percent of those who suffer from depression or bipolar disorder commit suicide each year. In 1996 the Centers for Disease Control and Prevention listed suicide as the ninth leading cause of death in the U.S. (slightly behind infection with the AIDS virus), taking the lives of 30,862 people. Most investigators, however, believe this number is a gross underestimate. Many people who kill themselves do so in a way that allows another diagnosis to be listed on the death certificate, so that families can receive insurance benefits or avoid embarrassment. Further, some fraction of automobile accidents unquestionably are concealed suicides.

The financial drain is enormous as well. In 1992 the estimated costs of depression totaled $43 billion, mostly from reduced or lost worker productivity.

Accumulating findings indicate that severe depression also heightens the risk of dying after a heart attack or stroke. And it often reduces the quality of life for cancer patients and might reduce survival time.

■ GENETIC FINDINGS

Geneticists have provided some of the oldest proof of a biological component to depression in many people. Depression and manic-depression frequently run in families. Thus, close blood relatives (children, siblings, and parents) of patients with severe depressive or bipolar disorder are much more likely to suffer from those or related conditions than are members of the general population. Studies of identical twins (who are genetically indistinguishable) and fraternal twins (whose genes generally are no more alike than those of other pairs of siblings) also support an inherited component. The finding of illness in both members of a pair is much higher for manic-depression in identical twins than in fraternal ones and is somewhat elevated for depression alone.

In the past 20 years, genetic researchers have expended great effort trying to identify the genes at fault. So far, though, those genes have evaded discovery, perhaps because a predisposition to depression involves several genes, each of which makes only a small, hard-to-detect contribution.

Preliminary reports from a study of an Amish population with an extensive history of manic-depression once raised the possibility that chromosome 11 held one or more genes producing vulnerability to bipolar disorder, but the finding did not hold up. A gene somewhere on the X chromosome could play a role in some cases of that condition, but the connection is not evident in most people who have been studied. Most recently, various regions of chromosome 18 and a site on chromosome 21 have been suggested to participate in vulnerability to bipolar illness, but these findings await replication.

As geneticists continue their searches, other investigators are concentrating on neurochemical aspects. Much of that work focuses on neurotransmitters. In particular, many cases of depression apparently stem at least in part from disturbances in brain circuits that convey signals through certain neurotransmitters of the monoamine class. These biochemicals, all derivatives of amino acids, include serotonin, norepinephrine, and dopamine; of these, only evidence relating to norepinephrine and serotonin is abundant.

Monoamines first drew the attention of depression researchers in the 1950s. Early in that decade, physicians discovered that severe depression arose in about 15 percent of patients who were treated for hypertension with the drug reserpine. This agent turned out to deplete monoamines. At about the same time doctors found that an agent prescribed against tuberculosis elevated mood in some users who were depressed. Follow-up investigations revealed that the drug inhibited the neuronal breakdown of monoamines by an enzyme (monoamine oxidase); presumably the agent eased depression by allowing monoamines to avoid degradation and to remain active in brain circuits. Together these findings implied that abnormally low levels of monoamines in the brain could cause depression. This insight led to the development of monoamine oxidase inhibitors as the first class of antidepressants.

■ THE NOREPINEPHRINE LINK

But which monoamines were most important in depression? In the 1960s Joseph J. Schildkraut of Harvard University cast his vote with norepinephrine in the now classic "catecholamine" hypothesis of mood disorders. He proposed that depression stems from a deficiency of norepinephrine (which is also classified as a catecholamine) in certain brain circuits and that mania arises from an overabundance of the substance. The theory has since been refined, acknowledging, for instance, that decreases or elevations in norepinephrine do not alter moods in everyone. Nevertheless, the proposed link between norepinephrine depletion and depression has gained much experimental support. These circuits originate in the brain stem, primarily in the pigmented locus coeruleus, and project to many areas of the brain, including to the limbic system—a group of cortical and subcortical areas that play a significant part in regulating emotions.

To understand the recent evidence relating to norepinephrine and other monoamines, it helps to know how those neurotransmitters work. The points of contact between two neurons, or nerve cells, are termed synapses. Monoamines, like all neurotransmitters, travel from one neuron (the presynaptic cell) across a small gap (the synaptic cleft) and attach to receptor molecules on the surface of the second neuron (the postsynaptic cell). Such binding elicits intracellular changes that stimulate or inhibit firing of the postsynaptic cell. The effect of the neurotransmitter depends greatly on the nature and concentration of its receptors on the postsynaptic cells. Serotonin receptors, for instance, come in 13 or more subtypes that can vary in their sensitivity to serotonin and in the effects they produce.

The strength of signaling can also be influenced by the amount of neurotransmitter released and by how long it remains in the synaptic cleft—properties influenced by at least two kinds of molecules on the surface of the releasing cells: autoreceptors and transporters. When an autoreceptor becomes bound by neurotransmitter molecules in the synapse, the receptors signal the cell to reduce its firing rate and thus its release of the transmitter. The transporters physically pump neurotransmitter molecules from the synaptic cleft back into presynaptic cells, a process termed reuptake. Monoamine oxidase inside cells can affect synaptic neurotransmitter levels as well, by degrading monoamines and so reducing the amounts of those molecules available for release [*see color plate 3*].

Among the findings linking impoverished synaptic norepinephrine levels to depression is the discovery in many studies that indirect markers of norepinephrine levels in the brain—levels of its metabolites, or by-products, in more accessible material (urine and cerebrospinal fluid)—are often low in depressed individuals. In addition, postmortem studies have revealed increased densities of certain norepinephrine receptors in the cortex of depressed suicide victims.

Observers unfamiliar with receptor display might assume that elevated numbers of receptors were a sign of more contact between norepinephrine and its receptors and more signal transmission. But this pattern of receptor "up-regulation" is actually one that scientists would expect if norepinephrine concentrations in synapses were abnormally low. When transmitter molecules become unusually scarce in synapses, postsynaptic cells often expand receptor numbers in a compensatory attempt to pick up whatever signals are available.

A recent discovery supporting the norepinephrine hypothesis is that new drugs selectively able to block norepinephrine reuptake, and so increase norepinephrine in synapses, are effective antidepressants in many people. One compound, reboxetine, is available as an antidepressant outside the U.S. and is awaiting approval here.

■ SEROTONIN CONNECTIONS

The data connecting norepinephrine to depression are solid and still growing. Yet research into serotonin has taken center stage in the 1990s, thanks to the therapeutic success of Prozac and related antidepressants that manipulate serotonin levels. Serious investigations into serotonin's role in mood disorders, however, have been going on for almost 30 years, ever since Arthur J. Prange, Jr., of the University of North Carolina at Chapel Hill, Alec Coppen of the Medical Research Council in England, and their co-workers put forward the so-called permissive hypothesis. This view held that synaptic depletion of serotonin was another cause of depression, one that worked by promoting, or "permitting," a fall in norepinephrine levels.

Defects in serotonin-using circuits could certainly dampen norepinephrine signaling. Serotonin-producing neurons project from the raphe nuclei in the brain stem to neurons in diverse regions of the central nervous system, including those that secrete or control the release of norepinephrine. Serotonin depletion might contribute to depression by affecting other kinds of neurons as well; serotonin-producing cells extend into many brain regions thought to participate in depressive symptoms—including the amygdala (an area involved in emotions), the hypothalamus (involved in appetite, libido, and sleep) and cortical areas that participate in cognition and other higher processes.

Among the findings supporting a link between low synaptic serotonin levels and depression is that cerebrospinal fluid in depressed, and especially in suicidal, patients contains reduced amounts of a major serotonin by-product (signifying reduced levels of serotonin in the brain itself). In addition, levels of a surface molecule unique to serotonin-releasing cells in the brain are lower in depressed patients than in healthy subjects, implying that the numbers of serotonergic cells are reduced. Moreover, the density of at least one form of serotonin receptor—type 2—is greater in postmortem brain tissue of depressed patients; as was true in studies of norepinephrine receptors, this up-regulation is suggestive of a compensatory response to too little serotonin in the synaptic cleft.

Further evidence comes from the remarkable therapeutic effectiveness of drugs that block presynaptic reuptake transporters from drawing serotonin out of the synaptic cleft. Tricyclic antidepressants (so-named because they contain three rings of chemical groups) joined monoamine oxidase inhibitors on pharmacy shelves in the late 1950s, although their mechanism of action was not known at the time.

Eventually, though, they were found to produce many effects in the brain, including a decrease in serotonin reuptake and a consequent rise in serotonin levels in synapses.

Investigators suspected that this last effect accounted for their antidepressant action, but confirmation awaited the introduction in the late 1980s of Prozac and then other drugs (Paxil, Zoloft, and Luvox) able to block serotonin reuptake transporters without affecting other brain monoamines. These selective serotonin reuptake inhibitors (SSRIs) have now revolutionized the treatment of depression, because they are highly effective and produce much milder side effects than older drugs do. Today even newer antidepressants, such as Effexor, block reuptake of both serotonin and norepinephrine.

Studies of serotonin have also offered new clues to why depressed individuals are more susceptible to heart attack and stroke. Activation and clumping of blood platelets (cell-like structures in blood) contribute to the formation of thrombi that can clog blood vessels and shut off blood flow to the heart and brain, thus damaging those organs. Work in my laboratory and elsewhere has shown that platelets of depressed people are particularly sensitive to activation signals, including, it seems, to those issued by serotonin, which amplifies platelet reactivity to other, stronger chemical stimuli. Further, the platelets of depressed patients bear reduced numbers of serotonin reuptake transporters. In other words, compared with the platelets of healthy people, those in depressed individuals probably are less able to soak up serotonin from their environment and thus to reduce their exposure to platelet-activation signals.

Disturbed functioning of serotonin or norepinephrine circuits, or both, contributes to depression in many people, but compelling work can equally claim that depression often involves dysregulation of brain circuits that control the activities of certain hormones. Indeed, hormonal alterations in depressed patients have long been evident.

■ HORMONAL ABNORMALITIES

The hypothalamus of the brain lies at the top of the hierarchy regulating hormone secretion. It manufactures and releases peptides (small chains of amino acids) that act on the pituitary, at the base of the brain, stimulating or inhibiting the pituitary's release of various hormones into the blood. These hormones—among them growth hormone, thyroid-stimulating hormone, and adrenocorticotropic hormone (ACTH)—control the release of other hormones from target glands. In addition to functioning outside the nervous system, the hormones released in response to pituitary hormones feed back to the pituitary and hypothalamus. There they deliver inhibitory signals that keep hormone manufacture from becoming excessive.

Depressed patients have repeatedly been demonstrated to show a blunted response to a number of substances that normally stimulate the release of growth hormone. They also display aberrant responses to the hypothalamic substance that normally induces secretion of thyroid-stimulating hormone from the pituitary. In addition, a common cause of nonresponse to antidepressants is the presence of previously undiagnosed thyroid insufficiency.

All these findings are intriguing, but so far the strongest case has been made for dysregulation of the hypothalamic-pituitary-adrenal (HPA) axis—the system that manages the body's response to stress. When a threat to physical or psychological well-being is detected, the hypothalamus amplifies production of corticotropin-releasing factor (CRF), which induces the pituitary to secrete ACTH. ACTH then instructs the adrenal gland atop each kidney to release cortisol. Together all the changes prepare the body to fight or flee and cause it to shut down activities that would distract from self-protection. For instance, cortisol enhances the delivery of fuel to muscles. At the same time, CRF depresses the appetite for food and sex and heightens alertness. Chronic activation of the HPA axis, however, may lay the ground for illness and, it appears, for depression.

As long ago as the late 1960s and early 1970s, several research groups reported increased activity in the HPA axis in unmedicated depressed patients, as evinced by raised levels of cortisol in urine, blood, and cerebrospinal fluid, as well as by other measures. Hundreds, perhaps even thousands, of subsequent studies have confirmed that substantial numbers of depressed patients—particularly those most severely affected—display HPA-axis hyperactivity. Indeed, the finding is surely the most replicated one in all of biological psychiatry.

Deeper investigation of the phenomenon has now revealed alterations at each level of the HPA axis in depressed patients. For instance, both the adrenal gland and the pituitary are enlarged, and the adrenal gland hypersecretes cortisol. But many researchers, including my colleagues and me at Emory University, have become persuaded that aberrations in CRF-producing neurons of the hypothalamus and elsewhere bear most of the responsibility for HPA-axis hyperactivity and the emergence of depressive symptoms.

Notably, study after study has shown CRF concentrations in cerebrospinal fluid to be elevated in depressed patients, compared with control subjects or individuals with other psychiatric disorders. This magnification of CRF levels is reduced by treatment with antidepressants and by effective electroconvulsive therapy. Further, postmortem brain tissue studies have revealed a marked exaggeration both in the number of CRF-producing neurons in the hypothalamus and in the expression of

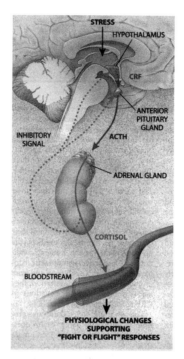

Hormonal system known as the hypothalamic-pituitary-adrenal axis is active (solid arrows) in response to stress. Mounting evidence suggests that chronic overactivity of the axis, and particularly overproduction of the corticotropin-releasing factor (CRF), contributes to depression. For instance, application of CRF to the brain of laboratory animals produces symptoms similar to those often displayed by depressed individuals. (Tomo Narashima)

the CRF gene (resulting in elevated CRF synthesis) in depressed patients as compared with controls. Moreover, delivery of CRF to the brains of laboratory animals produces behavioral effects that are cardinal features of depression in humans, namely, insomnia, decreased appetite, decreased libido, and anxiety.

Neurobiologists do not yet now exactly how the genetic, monoamine, and hormonal findings piece together, if indeed they always do. The discoveries nonetheless suggest a partial scenario for how people who endure traumatic childhoods become depressed later in life. I call this hypothesis the stress-diathesis model of mood disorders, in recognition of the interaction between experience (stress) and inborn predisposition (diathesis).

The observation that depression runs in families means that certain genetic traits in the affected families somehow lower the threshold for depression. Conceivably, the genetic features directly or indirectly diminish monoamine levels in synapses or increase reactivity of the HPA axis to stress. The genetically determined threshold is not necessarily low enough to induce depression in the absence of serious stress but may then be pushed still lower by early, adverse life experiences.

My colleagues and I propose that early abuse or neglect not only activates the stress response but induces persistently increased activity in CRF-containing neurons, which are known to be stress responsive and to be overactive in depressed people. If the hyperactivity in the neurons of children persisted through adulthood, these supersensitive cells would react vigorously even to mild stressors. This effect in people already innately predisposed to depression could then produce both the neuroendocrine and behavioral responses characteristic of the disorder.

■ SUPPORT FOR A MODEL

To test the stress-diathesis hypothesis, we have conducted a series of experiments in which neonatal rats were neglected. We removed them from their mothers for brief periods on about 10 of their first 21 days of life, before allowing them to grow up (after weaning) in a standard rat colony. As adults, these maternally deprived rats showed clear signs of changes in CRF-containing neurons, all in the direction observed in depressed patients—such as rises in stress-induced ACTH secretion and elevations of CRF concentrations in several areas of the brain. Levels of corticosterone (the rats's cortisol) also rose. These findings suggested that a permanent increase in CRF gene expression and thus in CRF production occurred in the maternally deprived rats, an effect now confirmed by Paul M. Plotsky, one of my coworkers at Emory.

We have also found an increase in CRF-receptor density in certain brain regions of maternally deprived rats. Receptor amplification commonly reflects an attempt to compensate for a decrease in the substance that acts on the receptor. In this case, though, the rise in receptor density evidently occurs not as a balance to decreased CRF but in spite of an increase—the worst of all possibilities. Permanently elevated receptor concentrations would tend to magnify the action of CRF, thereby forever enhancing the depression-inducing effects of CRF and stress.

In an exciting preliminary finding, Plotsky has observed that treatment with one of the selective serotonin reuptake inhibitors (Paxil) returns CRF levels to normal, compensates for any gain in receptor sensitivity or number (as indicated by normal corticosterone production lower down in the axis), and normalizes behavior (for instance, the rats become less fearful).

We do not know exactly how inhibition of serotonin reuptake would lead to normalization of the HPA axis. Even so, the finding implies that serotonin reuptake inhibitors might be particularly helpful in depressed patients with a history of childhood trauma. Plotsky further reports that all the HPA-axis and CRF abnormalities returned when treatment stopped, a hint that pharmaceutical therapy in analogous human patients might have to be continued indefinitely to block recurrences of depression.

Studies of Bonnet macaque monkeys, which as primates more closely resemble humans, yielded similar results. Newborns and their mothers encountered three foraging conditions for three months after the babies' birth: a plentiful, a scarce, and a variable food supply. The variable situation (in which food was available unpredictably) evoked considerable anxiety in monkey mothers, who became so anxious and preoccupied that they basically ignored their offspring. As our model predicts, the neonates in the variable-foraging condition were less active, withdrew from interactions with other monkeys, and froze in novel situations. In adulthood, they also exhibited marked elevations in CRF concentrations in spinal fluid.

The rat and monkey data raise profound clinical and public health questions. In the U.S. alone in 1995, more than three million children were reportedly abused or neglected, and at least a million of those reports were verified. If the effects in human beings resemble those of the animals, the findings imply that abuse or neglect may produce permanent changes in the developing brain—changes that chronically boost the output of, and responsiveness to, CRF, and therefore increase the victims' lifelong vulnerability to depression.

If that conclusion is correct, investigators will be eager to determine whether noninvasive techniques able to assess the activity of CRF-producing neurons or the number of CRF receptors could identify abused individuals at risk for later depression. In addition, they will want to evaluate whether antidepressants or other interventions, such as psychotherapy, could help prevent depression in children who are shown to be especially susceptible. Researchers will also need to find out whether depressed adults with a history of abuse need to take antidepressants in perpetuity and whether existing drugs or psychotherapy can restore normal activity in CRF-producing neurons in humans.

The stress-diathesis model does not account for all cases of depression; not everyone who is depressed has been neglected or abused in childhood. But individuals who have both a family history of the condition and a traumatic childhood seem to be unusually prone to the condition. People who have no genetic predisposition to depression (as indicated by no family history of the disorder) could conceivably be relatively protected from serious depression even if they have a bad childhood or severe trauma later in life. Conversely, some people who have a strong inherited vulnerability will find themselves battling depression even when their childhoods and later life are free of trauma.

More work on the neurobiology of depression is clearly indicated, but the advances achieved so far are already being translated into ideas for new medications. Several pharmaceutical houses are developing blockers of CRF receptors to test the antidepressant value of such agents. Another promising class of drugs activates specific serotonin receptors; such agents can potentially exert powerful antidepressive effects without stimulating serotonin receptors on neurons that play no part in depression.

More therapies based on new understandings of the biology of mood disorders are sure to follow as well. As research into the neurobiological underpinnings progresses, treatment should become ever more effective and less likely to produce unwanted side effects.

—June 1998

MANIC-DEPRESSIVE ILLNESS AND CREATIVITY

Does some fine madness plague great artists? Several studies now show that creativity and mood disorders are linked

Kay Redfield Jamison

"Men have called me mad," wrote Edgar Allan Poe, "but the question is not yet settled, whether madness is or is not the loftiest intelligence—whether much that is glorious—whether all that is profound—does not spring from disease of thought—from moods of mind exalted at the expense of the general intellect."

Many people have long shared Poe's suspicion that genius and insanity are entwined. Indeed, history holds countless examples of "that fine madness." Scores of influential 18th- and 19th-century poets, notably William Blake, Lord Byron, and Alfred, Lord Tennyson, wrote about the extreme mood swings they endured. Modern American poets John Berryman, Randall Jarrell, Robert Lowell, Sylvia Plath, Theodore Roethke, Delmore Schwartz, and Anne Sexton were all hospitalized for either mania or depression during their lives. And many painters and composers, among them Vincent van Gogh, Georgia O'Keeffe, Charles Mingus, and Robert Schumann, have been similarly afflicted.

Judging by current diagnostic criteria, it seems that most of these artists—and many others besides—suffered from one of the major mood disorders, namely, manic-depressive illness or major depression. Both are fairly common, very treatable, and yet frequently lethal diseases. Major depression induces intense melancholic spells, whereas manic-depression, a strongly genetic disease, pitches patients repeatedly from depressed to hyperactive and euphoric, or intensely irritable, states. In its milder form, termed cyclothymia, manic-depression causes pronounced but not totally debilitating changes in mood, behavior, sleep, thought patterns, and energy levels. Advanced cases are marked by dramatic, cyclic shifts.

Could such disruptive diseases convey certain creative advantages? Many people find that proposition counterintuitive. Most manic-depressives do not possess extraordinary imagination, and most accomplished artists do not suffer from recurring mood swings. To assume, then, that such diseases usually promote artistic talent strongly reinforces simplistic notions of the "mad genius." Worse yet, such a generalization trivializes a very serious medical condition and, to some degree, discredits individuality in the arts as well. It would be wrong to label anyone who is unusually accomplished, energetic, intense, moody, or eccentric as manic-depressive. All the same, recent studies indicate that a high number of established artists—far more than could be expected by chance—meet the diagnostic criteria of manic-depression or major depression given in the fourth edition of the *Diagnostic and Statistical Manual of Mental Disorders (DSM-IV)*. In fact, it seems that these diseases can sometimes enhance or otherwise contribute to creativity in some people.

By virtue of their prevalence alone, it is clear that mood disorders do not necessarily breed genius. Indeed, 1 percent of the general population suffer from manic-depression, also called bipolar disorder, and 5 percent from a major depression, or unipolar disorder, during their lifetime. Depression affects twice as many women as men and most often, but not always, strikes later in life. Bipolar disorder afflicts equal numbers of women and men, and more than a third of all cases surface before age 20. Some 60 to 80 percent of all adolescents and adults who commit suicide have a history of bipolar or unipolar illness. Before the late 1970s, when the drug lithium first became widely available, one person in five with manic-depression committed suicide.

Major depression in both unipolar and bipolar disorders manifests itself through apathy, lethargy, hopelessness, sleep disturbances, slowed physical movements and thinking, impaired memory and concentration, and a loss of pleasure in typically enjoyable events. The diagnostic criteria also include suicidal thinking, self-blame, and inappropriate guilt. To distinguish clinical depression from normal periods of unhappiness, the common guidelines further require that these symptoms persist for a minimum of two to four weeks and also that they significantly interfere with a person's everyday functioning.

■ MOOD ELEVATION

During episodes of mania or hypomania (mild mania), bipolar patients experience symptoms that are in many ways the opposite of those associated with depression. Their mood and self-esteem are elevated. They sleep less and have abundant energy; their productivity increases. Manics frequently become paranoid and irritable. Moreover, their speech is often rapid, excitable, and intrusive, and their thoughts move quickly and fluidly from one topic to another. They usually hold tremendous conviction about the correctness and importance of their own ideas as well. This grandiosity can contribute to poor judgment and impulsive behavior.

Hypomanics and manics generally have chaotic personal and professional relationships. They may spend large sums of money, drive recklessly, or pursue questionable business ventures or sexual liaisons. In some cases, manics suffer from violent agitation and delusional thoughts as well as visual and auditory hallucinations.

■ RATES OF MOOD DISORDERS

For years, scientists have documented some kind of connection between mania, depression, and creative output. In the late 19th and early 20th centuries, researchers turned to accounts of mood disorders written by prominent artists, their physicians, and friends. Although largely anecdotal, this work strongly suggested that renowned writers, artists, and composers—and their first-degree relatives—were far more likely to experience mood disorders and to commit suicide than was the general population. During the past 20 years, more systematic studies of artistic populations have confirmed these findings [*see illustration on page 280*]. Diagnostic and psychological analyses of living writers and artists can give quite meaningful estimates of the rates and types of psychopathology they experience.

In the 1970s Nancy C. Andreasen of the University of Iowa completed the first of these rigorous studies, which made use of structured interviews, matched control groups, and strict diagnostic criteria. She examined 30 creative writers and found an extraordinarily high occurrence of mood disorders and alcoholism among them. Eighty percent had experienced at least one episode of major depression, hypoma-

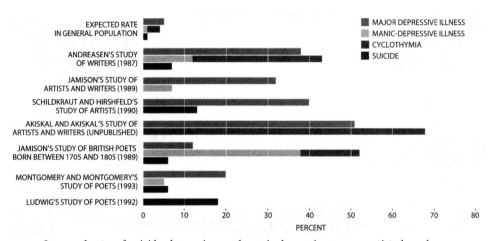

Increased rates of suicide, depression, and manic-depression among artists have been established by many separate studies. These investigations show that artists experience up to 18 times the rate of suicide seen in the general population, eight to 10 times the rate of depression, and 10 to 20 times the rate of manic-depression and its milder form, cyclothymia. (Lisa Burnett)

nia, or mania; 43 percent reported a history of hypomania or mania. Also, the relatives of these writers, compared with the relatives of the control subjects, generally performed more creative work and more often had a mood disorder.

A few years later, while on sabbatical in England from the University of California at Los Angeles, I began a study of 47 distinguished British writers and visual artists. To select the group as best I could for creativity, I purposefully chose painters and sculptors who were Royal Academicians or Associates of the Royal Academy. All the playwrights had won the New York Drama Critics Award or the Evening Standard Drama (London Critics) Award, or both. Half of the poets were already represented in the *Oxford Book of Twentieth Century English Verse*. I found that 38 percent of these artists and writers had in fact been previously treated for a mood disorder; three fourths of those treated had required medication or hospitalization, or both. And half of the poets—the largest fraction from any one group—had needed such extensive care.

Hagop S. Akiskal of the University of California at San Diego, also affiliated with the University of Tennessee at Memphis, and his wife, Kareen Akiskal, subsequently interviewed 20 award-winning European writers, poets, painters, and sculptors. Some two thirds of their subjects exhibited recurrent cyclothymic or hypomanic tendencies, and half had at one time suffered from a major depression. In collaboration with David H. Evans of the University of Memphis, the Akiskals noted the same trends among living blues musicians. More recently Stuart A. Montgomery and his wife, Deirdre B. Montgomery, of St. Mary's Hospital in London examined 50 modern British poets. One fourth met current diagnostic criteria for depression or manic-depression; suicide was six times more frequent in this community than in the general population.

Ruth L. Richards and her colleagues at Harvard University set up a system for assessing the degree of original thinking required to perform certain creative tasks. Then, rather than screening for mood disorders among those already deemed highly inventive, they attempted to rate creativity in a sample of manic-depressive patients. Based on their scale, they found that compared with individuals having no personal or family history of psychiatric disorders, manic-depressive and cyclothymic patients (as well as their unaffected relatives) showed greater creativity.

Biographical studies of earlier generations of artists and writers also show consistently high rates of suicide, depression, and manic-depression—up to 18 times the rate of suicide seen in the general population, eight to 10 times that of depression, and 10 to 20 times that of manic-depressive illness and its milder variants. Joseph J. Schildkraut and his co-workers at Harvard concluded that approximately half of the 15 20th-century abstract-expressionist artists they studied suffered from depressive or manic-depressive illness; the suicide rate in this group was at least 13 times the current U.S. national rate.

In 1992 Arnold M. Ludwig of the University of Kentucky published an extensive biographical survey of 1,005 famous 20th-century artists, writers, and other professionals, some of whom had been in treatment for a mood disorder. He discovered that the artists and writers experienced two to three times the rate of psychosis, suicide attempts, mood disorders, and substance abuse that comparably successful people in business, science, and public life did. The poets in this sample had most often been manic or psychotic and hospitalized; they also proved to be some 18 times more likely to commit suicide than is the general public. In a comprehensive biographical study of 36 major British poets born between 1705 and 1805, I found similarly elevated rates of psychosis and severe psychopathology. These poets were 30 times more likely to have had manic-depressive illness than were their contemporaries, at least 20 times more likely to have been committed to an asylum, and some five times more likely to have taken their own life.

These corroborative studies have confirmed that highly creative individuals experience major mood disorders more often than do other groups in the general population. But what does this mean for their work? How does a psychiatric illness contribute to creative achievement? First, the common features of hypomania seem highly conducive to original thinking; the diagnostic criteria for this phase of the disorder include "sharpened and unusually creative thinking and increased productivity." And accumulating evidence suggests that the cognitive styles associated with hypomania (expansive thought and grandiose moods) can lead to increased fluency and frequency of thoughts.

■ MANIA AND CREATIVITY

Studying the speech of hypomanic patients has revealed that they tend to rhyme and use other sound associations, such as alliteration, far more often than do unaffected individuals. They also use idiosyncratic words nearly three times as often as do control subjects. Moreover, in specific drills, they can list synonyms or form other word associations much more rapidly than is considered normal. It seems, then, that both quantity and quality of thoughts build during hypomania. This speed increase may range from a very mild quickening to complete psychotic incoherence. It is not yet clear what causes this qualitative change in mental processing. Nevertheless, this altered cognitive state may well facilitate the formation of unique ideas and associations.

People with manic-depressive illness and those who are creatively accomplished share certain noncognitive features: the ability to function well on a few hours of sleep, the focus needed to work intensively, bold and restless attitudes, and an ability to experience a profound depth and variety of emotions. The less dramatic daily aspects of manic-depression might also provide creative advantage to some individuals. The manic-depressive temperament is, in a biological sense, an alert, sensitive system that reacts strongly and swiftly. It responds to the world with a wide range of emotional, perceptual, intellectual, behavioral, and energy changes. In a sense, depression is a view of the world through a dark glass, and mania is that seen through a kaleidoscope—often brilliant but fractured.

Where depression questions, ruminates, and hesitates, mania answers with vigor and certainty. The constant transitions in and out of constricted and then expansive thoughts, subdued and then violent responses, grim and then ebullient moods, withdrawn and then outgoing stances, cold and then fiery states—and the rapidity and fluidity of moves through such contrasting experiences—can be painful and confusing. Ideally, though, such chaos in those able to transcend it or shape it to their will can provide a familiarity with transitions that is probably useful in artistic endeavors. This vantage readily accepts ambiguities and the counteracting forces in nature.

Extreme changes in mood exaggerate the normal tendency to have conflicting selves; the undulating, rhythmic, and transitional moods and cognitive changes so characteristic of manic-depressive illness can blend or harness seemingly contradic-

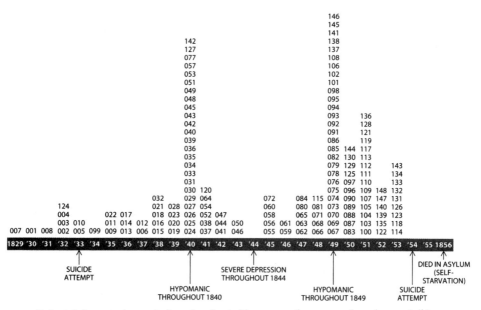

Robert Schumann's musical works, charted by year and opus number, show a striking relation between his mood states and his productivity. He composed the most when hypomanic and the least when depressed. Both of Schumann's parents were clinically depressed, and two other first-degree relatives committed suicide. Schumann himself attempted suicide twice and died in an insane asylum. One of his sons spent more than 30 years in a mental institution. (Adapted from E. Slater and A. Meyer, 1999)

tory moods, observations, and perceptions. Ultimately, these fluxes and yokings may reflect truth in humanity and nature more accurately than could a more fixed viewpoint. The "consistent attitude toward life" may not, as Byron scholar Jerome J. McGann of the University of Virginia points out, be as insightful as an ability to live with, and portray, constant change.

The ethical and societal implications of the association between mood disorders and creativity are important but poorly understood. Some treatment strategies pay insufficient heed to the benefits manic-depressive illness can bestow on some individuals. Certainly most manic-depressives seek relief from the disease, and lithium and anticonvulsant drugs are very effective therapies for manias and depressions. Nevertheless, these drugs can dampen a person's general intellect and limit his or her emotional and perceptual range. For this reason, many manic-depressive patients stop taking these medications.

Left untreated, however, manic-depressive illness often worsens over time—and no one is creative when severely depressed, psychotic, or dead. The attacks of both mania and depression tend to grow more frequent and more severe. Without regular treatment the disease eventually becomes less responsive to medication. In addition, bipolar and unipolar patients frequently abuse mood-altering substances, such as alcohol and illicit drugs, which can cause secondary medical and emotional burdens for manic-depressive and depressed patients.

■ THE GOAL OF TREATMENT

The real task of imaginative, compassionate, and effective treatment, therefore, is to give patients more meaningful choices than they are now afforded. Useful intervention must control the extremes of depression and psychosis without sacrificing crucial human emotions and experiences. Given time and increasingly sophisticated research, psychiatrists will likely gain a better understanding of the complex biological basis for mood disorders. Eventually, the development of new drugs should make it possible to treat manic-depressive individuals so that those aspects of temperament and cognition that are essential to the creative process remain intact.

The development of more specific and less problematic therapies should be swift once scientists find the gene, or genes, responsible for the disease. Prenatal tests and other diagnostic measures may then become available; these possibilities raise a host of complicated ethical issues. It would be irresponsible to romanticize such a painful, destructive, and all too often deadly disease. Hence, 3 to 5 percent of the Human Genome Project's total budget (which is conservatively estimated at $3 billion) has been set aside for studies of the social, ethical, and legal implications of genetic research. It is hoped that these investigations will examine the troubling issues surrounding manic-depression and major depression at length. To help those who have manic-depressive illness, or who are at risk for it, must be a major public health priority.

—February 1995

Consciousness

THE PUZZLE OF CONSCIOUS EXPERIENCE

Neuroscientists and others are at last plumbing one of the most profound mysteries of existence. But knowledge of the brain alone may not get them to the bottom of it

David J. Chalmers

Conscious experience is at once the most familiar thing in the world and the most mysterious. There is nothing we know about more directly than consciousness, but it is extraordinarily hard to reconcile it with everything else we know. Why does it exist? What does it do? How could it possibly arise from neural processes in the brain? These questions are among the most intriguing in all of science.

From an objective viewpoint, the brain is relatively comprehensible. When you look at this page, there is a whir of processing: photons strike your retina, electrical signals are passed up your optic nerve and between different areas of your brain, and eventually you might respond with a smile, a perplexed frown, or a remark. But there is also a subjective aspect. When you look at the page, you are conscious of it, directly experiencing the images and words as part of your private, mental life. You have vivid impressions of the colors and shapes of the images. At the same time, you may be feeling some emotions and forming some thoughts. Together such experiences make up consciousness: the subjective, inner life of the mind.

For many years, consciousness was shunned by researchers studying the brain and the mind. The prevailing view was that science, which depends on objectivity, could not accommodate something as subjective as consciousness. The behaviorist movement in psychology, dominant earlier in this century, concentrated on external behavior and disallowed any talk of internal mental processes. Later, the rise of cognitive science focused attention on processes inside the head. Still, consciousness remained off-limits, fit only for late-night discussion over drinks.

Over the past several years, however, an increasing number of neuroscientists, psychologists, and philosophers have been rejecting the idea that consciousness cannot be studied and are attempting to delve into its secrets. As might be expected of a field so new, there is a tangle of diverse and conflicting theories, often using basic concepts in incompatible ways. To help unsnarl the tangle, philosophical reasoning is vital.

The myriad views within the field range from reductionist theories, according to which consciousness can be explained by the standard methods of neuroscience and psychology, to the position of the so-called mysterians, who say we will never understand consciousness at all. I believe that on close analysis both of these views can be seen to be mistaken and that the truth lies somewhere in the middle.

Against reductionism I will argue that the tools of neuroscience cannot provide a full account of conscious experience, although they have much to offer. Against mysterianism I will hold that consciousness might be explained by a new kind of theory. The full details of such a theory are still out of reach, but careful reasoning and some educated inferences can reveal something of its general nature. For example, it will probably involve new fundamental laws, and the concept of information may play a central role. These faint glimmerings suggest that a theory of consciousness may have startling consequences for our view of the universe and of ourselves.

■ THE HARD PROBLEM

Researchers use the word "consciousness" in many different ways. To clarify the issues, we first have to separate the problems that are often clustered together under the name. For this purpose, I find it useful to distinguish between the "easy problems" and the "hard problem" of consciousness. The easy problems are by no means trivial—they are actually as challenging as most in psychology and biology—but it is with the hard problem that the central mystery lies.

The easy problems of consciousness include the following: How can a human subject discriminate sensory stimuli and react to them appropriately? How does the brain integrate information from many different sources and use this information to control behavior? How is it that subjects can verbalize their internal states? Although all these questions are associated with consciousness, they all concern the objective mechanisms of the cognitive system. Consequently, we have every reason to expect that continued work in cognitive psychology and neuroscience will answer them.

The hard problem, in contrast, is the question of how physical processes in the brain give rise to subjective experience. This puzzle involves the inner aspect of thought and perception: the way things feel for the subject. When we see, for example, we experience visual sensations, such as that of vivid blue. Or think of the ineffable sound of a distant oboe, the agony of an intense pain, the sparkle of happiness, or the meditative quality of a moment lost in thought. All are part of what I call consciousness. It is these phenomena that pose the real mystery of the mind.

To illustrate the distinction, consider a thought experiment devised by the Australian philosopher Frank Jackson. Suppose that Mary, a neuroscientist in the 23rd century, is the world's leading expert on the brain processes responsible for color vision. But Mary has lived her whole life in a black-and-white room and has never seen any other colors. She knows everything there is to know about physical processes in the brain—its biology, structure, and function. This understanding enables her to grasp all there is to know about the easy problems: how the brain discriminates stimuli, integrates information, and produces verbal reports. From her knowledge of color vision, she knows how color names correspond with wavelengths on the light spectrum. But there is still something crucial about color vision that Mary does not know: what it is like to experience a color such as red. It follows that there are facts about conscious experience that cannot be deduced from physical facts about the functioning of the brain.

Indeed, nobody knows why these physical processes are accompanied by conscious experience at all. Why is it that when our brains process light of a certain wavelength, we have an experience of deep purple? Why do we have any experience at all? Could not an unconscious automaton have performed the same tasks just as well? These are questions that we would like a theory of consciousness to answer.

IS NEUROSCIENCE ENOUGH?

I am not denying that consciousness arises from the brain. We know, for example, that the subjective experience of vision is closely linked to processes in the visual cortex. It is the link itself that perplexes, however. Remarkably, subjective experience seems to emerge from a physical process. But we have no idea how or why this is.

Given the flurry of recent work on consciousness in neuroscience and psychology, one might think this mystery is starting to be cleared up. On closer examination, however, it turns out that almost all the current work addresses only the easy problems of consciousness. The confidence of the reductionist view comes from the progress on the easy problems, but none of this makes any difference where the hard problem is concerned.

Consider the hypothesis put forward by neurobiologists Francis Crick of the Salk Institute for Biological Studies in San Diego and Christof Koch of the California Institute of Technology. They suggest that consciousness may arise from certain oscillations in the cerebral cortex, which become synchronized as neurons fire 40 times per second. Crick and Koch believe the phenomenon might explain how different attributes of a single perceived object (its color and shape, for example), which are processed in different parts of the brain are merged into a coherent whole. In this theory, two pieces of information become bound together precisely when they are represented by synchronized neural firings.

The hypothesis could conceivably elucidate one of the easy problems about how information is integrated in the brain. But why should synchronized oscillations give rise to a visual experience, no matter how much integration is taking place? This question involves the hard problem, about which the theory has nothing to offer. Indeed, Crick and Koch are agnostic about whether the hard problem can be solved by science at all.

The same kind of critique could be applied to almost all the recent work on consciousness. In his 1991 book *Consciousness Explained,* philosopher Daniel C. Dennett laid out a sophisticated theory of how numerous independent processes in the brain combine to produce a coherent response to a perceived event. The theory might do much to explain how we produce verbal reports on our internal states, but it tells us very little about why there should be a subjective experience behind these reports. Like other reductionist theories, Dennett's is a theory of the easy problems.

The critical common trait among these easy problems is that they all concern how a cognitive or behavioral function is performed. All are ultimately questions about how the brain carries out some task—how it discriminates stimuli, integrates information, produces reports, and so on. Once neurobiology specifies appropriate neural mechanisms, showing how the functions are performed, the easy problems are solved.

The hard problem of consciousness, in contrast, goes beyond problems about how functions are performed. Even if every behavioral and cognitive function related to consciousness were explained, there would still remain a further mystery:

Why is the performance of these functions accompanied by conscious experience? It is this additional conundrum that makes the hard problem hard.

■ THE EXPLANATORY GAP

Some have suggested that to solve the hard problem, we need to bring in new tools of physical explanation: nonlinear dynamics, say, or new discoveries in neuroscience, or quantum mechanics. But these ideas suffer from exactly the same difficulty. Consider a proposal from Stuart R. Hameroff of the University of Arizona and Roger Penrose of the University of Oxford. They hold that consciousness arises from quantum-physical processes taking place in microtubules, which are protein structures inside neurons. It is possible (if not likely) that such a hypothesis will lead to an explanation of how the brain makes decisions or even how it proves mathematical theorems, as Hameroff and Penrose suggest. But even if it does, the theory is silent about how these processes might give rise to conscious experience. Indeed, the same problem arises with any theory of consciousness based only on physical processing.

The trouble is that physical theories are best suited to explaining why systems have a certain physical structure and how they perform various functions. Most problems in science have this form; to explain life, for example, we need to describe how a physical system can reproduce, adapt, and metabolize. But consciousness is a different sort of problem entirely, as it goes beyond the scientific explanation of structure and function.

Of course, neuroscience is not irrelevant to the study of consciousness. For one, it may be able to reveal the nature of the neural correlate of consciousness—the brain processes most directly associated with conscious experience. It may even give a detailed correspondence between specific processes in the brain and related components of experience. But until we know why these processes give rise to conscious experience at all, we will not have crossed what philosopher Joseph Levine has called the explanatory gap between physical processes and consciousness. Making that leap will demand a new kind of theory.

In searching for an alternative, a key observation is that not all entities in science are explained in terms of more basic entities. In physics, for example, space-time, mass, and charge (among other things) are regarded as fundamental features of the world, as they are not reducible to anything simpler. Despite this irreducibility, detailed and useful theories relate these entities to one another in terms of fundamental laws. Together these features and laws explain a great variety of complex and subtle phenomena.

It is widely believed that physics provides a complete catalogue of the universe's fundamental features and laws. As physicist Steven Weinberg puts it in his 1992 book *Dreams of a Final Theory,* the goal of physics is a "theory of everything" from which all there is to know about the universe can be derived. But Weinberg concedes that there is a problem with consciousness. Despite the power of physical theory, the existence of consciousness does not seem to be derivable from physical laws. He defends physics by arguing that it might eventually explain what he calls the objective correlates of consciousness (that is, the neural correlates), but of course to do this is not to explain consciousness itself. If the existence of consciousness cannot be derived from physical laws, a theory of physics is not a true theory of everything. So a final theory must contain an additional fundamental component.

■ A TRUE THEORY OF EVERYTHING

Toward this end, I propose that conscious experience be considered a fundamental feature, irreducible to anything more basic. The idea may seem strange at first, but consistency seems to demand it. In the 19th century it turned out that electromagnetic phenomena could not be explained in terms of previously known principles. As a consequence, scientists introduced electromagnetic charge as a new fundamental entity and studied the associated fundamental laws. Similar reasoning should be applied to consciousness. If existing fundamental theories cannot encompass it, then something new is required.

Where there is a fundamental property, there are fundamental laws. In this case, the laws must relate experience to elements of physical theory. These laws will almost certainly not interfere with those of the physical world; it seems that the latter form a closed system in their own right. Rather the laws will serve as a bridge, specifying how experience depends on underlying physical processes. It is this bridge that will cross the explanatory gap.

Thus, a complete theory will have two components: physical laws, telling us about the behavior of physical systems from the infinitesimal to the cosmological, and what we might call psychophysical laws, telling us how some of those systems are associated with conscious experience. These two components will constitute a true theory of everything.

Supposing for the moment that they exist, how might we uncover such psychophysical laws? The greatest hindrance in this pursuit will be a lack of data. As I have described it, consciousness is subjective, so there is no direct way to monitor it in others. But this difficulty is an obstacle, not a dead end. For a start, each one of us has access to our own experiences, a rich trove that can be used to formulate theories. We can also plausibly rely on indirect information, such as subjects' descriptions of their experiences. Philosophical arguments and thought experiments also have a role to play. Such methods have limitations, but they give us more than enough to get started.

These theories will not be conclusively testable, so they will inevitably be more speculative than those of more conventional scientific disciplines. Nevertheless, there is no reason they should not be strongly constrained to account accurately for our own first-person experiences, as well as the evidence from subjects' reports. If we find a theory that fits the data better than any other theory of equal simplicity, we will have good reason to accept it. Right now we do not have even a single theory that fits the data, so worries about testability are premature.

We might start by looking for high-level bridging laws, connecting physical processes to experience at an everyday level. The basic contour of such a law might be gleaned from the observation that when we are conscious of something, we are generally able to act on it and speak about it—which are objective, physical functions. Conversely, when some information is directly available for action and speech, it is generally conscious. Thus, consciousness correlates well with what we might call "awareness": the process by which information in the brain is made globally available to motor processes such as speech and bodily action.

■ OBJECTIVE AWARENESS

The notion may seem trivial. But as defined here, awareness is objective and physical, whereas consciousness is not. Some refinements to the definition of awareness are needed, in order to extend the concept to animals and infants, which cannot speak. But at least in familiar cases, it is possible to see the rough outlines of a psychological law: where there is awareness, there is consciousness, and vice versa.

To take this line of reasoning a step further, consider the structure present in the conscious experience. The experience of a field of vision, for example, is a constantly changing mosaic of colors, shapes, and patterns and as such has a detailed geometric structure. The fact that we can describe this structure, reach out in the direction of many of its components, and perform other actions that depend on it suggests that the structure corresponds directly to that of the information made available in the brain through the neural processes of objective awareness.

Similarly, our experiences of color have an intrinsic three-dimensional structure that is mirrored in the structure of information processes in the brain's visual cortex. This structure is illustrated in the color wheels and charts used by artists. Colors are arranged in a systematic pattern—red to green on one axis, blue to yellow on another, and black to white on a third. Colors that are close to one another on a color wheel are experienced as similar. It is extremely likely that they also correspond to similar perceptual representations in the brain, as part of a system of complex three-dimensional coding among neurons that is not yet fully understood. We can recast the underlying concept as a principle of structural coherence: the structure of conscious experience is mirrored by the structure of information in awareness, and vice versa.

Another candidate for a psychophysical law is a principle of organizational invariance. It holds that physical systems with the same abstract organization will give rise to the same kind of conscious experience, no matter what they are made of. For example, if the precise interactions between our neurons could be duplicated with silicon chips, the same conscious experience would arise. The idea is somewhat controversial, but I believe it is strongly supported by thought experiments describing the gradual replacement of neurons by silicon chips. The remarkable implication is that consciousness might someday be achieved in machines.

The ultimate goal of a theory of consciousness is a simple and elegant set of fundamental laws, analogous to the fundamental laws of physics. The principles described above are unlikely to be fundamental, however. Rather they seem to be high-level psychophysical laws, analogous to macroscopic principles in physics such as those of thermodynamics or kinematics. What might the underlying fundamental laws be? No one knows, but I don't mind speculating.

I suggest that the primary psychophysical laws may centrally involve the concept of information. The abstract notion of information, as put forward in the 1940s by Claude E. Shannon of the Massachusetts Institute of Technology, is that of a set of separate states with a basic structure of similarities and differences between them. We can think of a 10-bit binary code as an information state, for example. Such information states can be embodied in the physical world. This happens whenever they correspond to physical states (voltages, say), and when differences between them can be transmitted along some pathway, such as a telephone line.

■ INFORMATION: PHYSICAL AND EXPERIENTIAL

We can also find information embodied in conscious experience. The pattern of color patches in a visual field, for example, can be seen as analogous to that of the pixels covering a display screen. Intriguingly, it turns out that we find the same information states embedded in conscious experience and in underlying physical processes in the brain. The three-dimensional encoding of color spaces, for example, suggests that the information state in a color experience corresponds directly to an information state in the brain. Thus, we might even regard the two states as distinct aspects of a single information state, which is simultaneously embodied in both physical processing and conscious experience.

A natural hypothesis ensues. Perhaps information, or at least some information, has two basic aspects: a physical one and an experiential one. This hypothesis has the status of a fundamental principle that might underlie the relation between physical process and experience. Wherever we find conscious experience, it exists as one aspect of an information state, the other aspect of which is embedded in a physical process in the brain. This proposal needs to be fleshed out to make a satisfying theory. But it fits nicely with the principles mentioned earlier—systems with the same organization will embody the same information, for example—and it could explain numerous features of our conscious experience.

The idea is at least compatible with several others, such as physicist John A. Wheeler's suggestion that information is fundamental to the physics of the universe. The laws of physics might ultimately be cast in informational terms, in which case we would have a satisfying congruence between the constructs in both physical and psychophysical laws. It may even be that a theory of physics and a theory of consciousness could eventually be consolidated into a single grander theory of information.

■ IS EXPERIENCE UBIQUITOUS?

A potential problem is posed by the ubiquity of information. Even a thermostat embodies some information, for example, but is it conscious? There are at least two possible responses. First, we could constrain the fundamental laws so that only some information has an experiential aspect, perhaps depending on how it is physically processed. Second, we might bite the bullet and allow that all information has an experiential aspect—where there is complex information processing, there is complex experience, and where there is simple information processing, there is simple experience. If this is so, then even a thermostat might have experiences, although they would be much simpler than even a basic color experience, and there would certainly be no accompanying emotions or thoughts. This seems odd at first, but if experience is truly fundamental, we might expect it to be widespread. In any case, the choice between these alternatives should depend on which can be integrated into the most powerful theory.

Of course, such ideas may be all wrong. On the other hand, they might evolve into a more powerful proposal that predicts the precise structure of our conscious experience from physical processes in our brains. If this project succeeds, we will have good reason to accept the theory. If it fails, other avenues will be pursued, and alternative fundamental theories may be developed. In this way, we may one day resolve the greatest mystery of the mind.

—December 1995

CAN SCIENCE EXPLAIN CONSCIOUSNESS?

What is consciousness? Can neurobiology explain it, or—
as some philosophers argue—does this most elusive and
inescapable of all phenomena lie beyond experiment's reach?

John Horgan

What was once the greatest mystery of biology, the human brain, is gradually yield-ing its secrets. Investigators are probing its deepest recesses with increasingly pow-erful tools, ranging from microelectrodes, which can discern the squeaks of indi-vidual neurons, to magnetic resonance imaging and positron emission tomography, which can amplify the cortical symphony evoked by, say, viewing Georges Seurat's *La Grande Jatte* or sniffing barbecued spareribs. With these and other techniques, re-searchers have begun elucidating the physiological processes underlying such facets of the mind as memory, perception, learning, and language.

Emboldened by these achievements, a growing number of scientists have dared to address what is simultaneously the most elusive and inescapable of all phenomena: consciousness, our immediate, subjective awareness of the world and of ourselves. Francis Crick should receive much of the credit—or blame—for the trend. Crick, who shared a Nobel Prize for the discovery of DNA's structure in 1953, turned to neuroscience shortly after he moved from England to the Salk Institute for Biological Studies in San Diego almost 20 years ago. Just as only the late Richard M. Nixon, famous for his red-baiting, could reestablish relations with communist China, so only Crick, who possesses a notoriously hard nose, could make consciousness a legitimate subject for science.

In 1990 Crick and Christof Koch, a young neuroscientist at the California Institute of Technology who collaborates closely with Crick, proclaimed in *Seminars in the Neurosciences* that the time was ripe for an assault on consciousness. They rejected the belief of many of their colleagues that consciousness cannot be defined, let alone studied. Consciousness, they argued, is really synonymous with awareness, and all forms of awareness—whether involving objects in the external world or highly abstract, internal concepts—seem to involve the same underlying mechanism, one that combines attention with short-term memory.

Contrary to the assumptions of cognitive scientists, philosophers, and others, Crick and Koch asserted, one cannot hope to achieve true understanding of consciousness or any other mental phenomenon by treating the brain as a black box—that is, an object whose internal structure is unknown and even irrelevant. Only by examining neurons and the interactions between them could scientists accumulate the kind of empirical, unambiguous knowledge that is required to create truly scientific models of consciousness, models analogous to those that explain transmission of genetic information by means of DNA.

Crick and Koch urged that investigators focus on visual awareness, since the visual system has already been well mapped in both animals and humans. If researchers could find the neural mechanisms underlying this function, they might unravel more complex and subtle phenomena, such as self-awareness, that may be unique to humans (and therefore much more difficult to study at the neural level). We may even comprehend why we have the paradoxical sensation of free will, an ineradicable sense that our minds exist independently of and exert control over our bodies. Crick elaborates on these ideas in *The Astonishing Hypothesis,* a book dedicated to Koch.

Crick's exhortations have helped incite an intellectual stampede in which mainstream researchers jostle with philosophers, computer scientists, psychiatrists, and other distinctly exotic species thirsting for insights into the mind. Meetings are proliferating. New publications have sprung up to feed the burgeoning interest, including *Psyche,* an E-mail journal based in Australia, and *Journal of Consciousness Studies,* a British quarterly.

Of course, neuroscientists are still far from agreeing on how consciousness should be studied or even defined. One prominent worker claims to have already "solved" consciousness: Gerald M. Edelman of the Scripps Research Institute, who shared a Nobel Prize in 1972 for research on antibodies. Edelman contends that our sense of awareness stems from a process he calls neural Darwinism, in which groups of neurons compete with one another to create an effective representation of the world. Edelman has promulgated this theory in a series of books—most recently, *Bright Air, Brilliant Fire,* published in 1992.

Crick has accused Edelman of dressing up ideas that are not terribly original in an idiosyncratic and obscure jargon. Most neuroscientists agree with this assessment (and find ludicrous the suggestion in a *New Yorker* magazine profile that Edelman might garner a second Nobel Prize for this work). But even those who admire Crick's efforts suspect his outlook might be too narrow. Gerald D. Fischbach of Harvard University, a former president of the Society for Neuroscience, says that it is not clear whether the kind of "electrophysiological" theory called for by Crick would suffice to explain consciousness, in the sense that discovering the structure of DNA accounted for heredity. "I don't think the field is mature enough to answer that question yet," he says.

■ THE NEW MYSTERIANS

Tomaso Poggio of the Massachusetts Institute of Technology, an authority on perception who was Koch's thesis adviser, thinks Crick may place too much emphasis on mechanisms that might coordinate, or bind together, the firings of neurons responding to a visual scene. Conversely, Crick may unduly neglect the role that the brain's plasticity, or ability to change its circuitry, might play in creating consciousness and other aspects of the mind, according to Poggio. Antonio R. Damasio of the University of Iowa, who maps our mental faculties by studying brain-damaged patients, holds that because a theory of consciousness must show how each of us acquires a sense of self, it must take into account not just the brain but the entire body. Damasio also believes that because consciousness is shaped by an individual's interactions with the environment and with other people, a neural model of consciousness will probably have to be supplemented by cognitive and social theories.

As neuroscientists debate these issues among themselves, others have challenged whether conventional neuroscience—despite its success in illuminating other attributes of the mind—can ever account for consciousness. Members of this eclectic group hail primarily from traditions outside mainstream neuroscience, such as physics and philosophy. Such individuals often seem less interested in clarifying consciousness than in mystifying it. For that reason, Owen Flanagan, a philosopher at Duke University, has dubbed some doubters "the new mysterians" (after the 1960s rock group Question Mark and the Mysterians, who performed the hit song "96 Tears").

One contingent of mysterians, whose most prominent member is Roger Penrose, a physicist at the University of Oxford, proposes that the mysteries of the mind must be related to the mysteries of quantum mechanics, which generates nondeterministic effects that classical theories of physics (and neuroscience) cannot. Although at first ignored and then derided by conventional neuroscientist, this alternative has steadily won popular attention through Penrose's efforts.

Another group of mysterians, which consists for the most part of philosophers, doubts whether any theory based on strictly materialistic effects—quantum or classical—can truly explain how and why we humans have a subjective experience of the world. "The question is, how can any physical system have a conscious state?" says Jerry A. Fodor, a philosopher at Rutgers University. Scientists who think science alone can answer that question "don't really understand it," Fodor declares.

None of these philosophers advocates dualism, a philosophy that holds that the mind exists independently of and can influence matter. But they reject the hardcore materialism of Crick, who claims in his new book that "your joys and your sorrows, your memories and your ambitions, your sense of personal identity and free will, are in fact no more than the behavior of a vast assembly of nerve cells and their associated molecules." Such a framework, they say, is inadequate for comprehending mental phenomena. Some other theory is required to make the relation between matter and mind "transparent," as the philosopher Thomas Nagel of New York University has put it.

Both these philosophical views and the quantum-consciousness theories are rejected by Terrence J. Sejnowski, a neural-network researcher at the Salk Institute. "I call these arguments from ignorance," says Sejnowski, who regularly shares tea with Crick and applauds his efforts to make consciousness a scientific subject. Sejnowski thinks that by following Crick's strict program neuroscientists "might actually get somewhere" in addressing biology's most profound riddle. Pointing out that life once seemed impossibly complex—before the discovery of DNA's structure revealed how information is passed from one generation to another—Sejnowski argues that much of the mystery veiling the mind will evaporate once scientists learn more about how the brain works.

■ "JUST LIKE WOODSTOCK"

Still, Sejnowski expresses the hope that the challenges from the philosophical skeptics and the quantum-consciousness claque, while misguided, may spur neuroscientists to be more creative as well as rigorous in their own research on consciousness. "When you get these wild disagreements," he says, "you also have an opportunity."

By that criterion, there should have been opportunities galore at the recent meeting on consciousness at the University of Arizona, billed as "the first interdisciplinary scientific conference on consciousness." "Wow, it's just like Woodstock," one speaker exclaims as he scans an auditorium crammed with neuroscientists, philosophers, psychiatrists, cognitive scientists, and others who defy categorization. Culture clashes abound. Indeed, the meeting offers a snapshot of a field—call it consciousness studies—in the throes of creation.

Crick is not here, but Koch is. So is Steen Rasmussen, a biochemist and computer scientist from the Santa Fe Institute, headquarters of the trendy fields of chaos and complexity. He suggests that the mind may be an "emergent"—that is, unpredictable and irreducible—property of the brain's complex behavior, just as James Joyce's *Ulysses* is a surprising outcome of applying the rules of spelling and grammar to the alphabet.

Other investigators offer more radical takes on consciousness. Brian D. Josephson of the University of Cambridge, who won a Nobel Prize in 1973 for discovering a subtle quantum effect that now bears his name, calls for a unified field theory that can account for mystical and even psychic experiences. Andrew T. Weil, a physician at the University of Arizona who is an authority on psychedelia, asserts that a complete theory of mind must address the reported ability of South American Indians who have ingested psychedelic drugs to experience identical hallucinations.

Just when one thinks that one has encountered every possible "paradigm" (a term much abused at this conference), another rears its head. After a speaker describes human thoughts as "quantum fluctuations of the vacuum energy of the universe"— which "is really God," she assures her listeners—an audience member stands to proclaim that physicists are discovering profound links between information, thermodynamics, and black holes, and these findings may help unlock the mysteries of consciousness. "There's not a black hole in our brains," he adds, "but—" "I think there *is* a black hole in my brain!" an overloaded listener interrupts.

Those who find lectures in the auditorium too staid can forage for more nonlinear fare in the hallway outside. "That's where all the really interesting stuff is," confides Spiros Antonopoulos, a journalist who sports a nose ring and chin braid and is covering the meeting for the cyberpunk magazine *Fringeware Review.* There one can join arguments over whether only humans are conscious or whether computers, bats, or even paramecia share this trait.

The conference does not lack empirical findings. Among the most intriguing are those involving patients whose sense of awareness has been damaged by some trauma or disease. Several investigators have studied patients who display the strange condition known as blindsight: they respond to visual stimuli—even catching a ball tossed to them, for example—while insisting that they cannot see anything.

Victor W. Mark, a neurologist from the University of North Dakota, shows a videotape of a young woman suffering from epilepsy so debilitating that to relieve it surgeons severed her corpus callosum, the bundle of neurons linking the two hemispheres of her brain. Although the operation alleviated her seizures, she was left with two centers of consciousness vying for dominance. When asked if her left hand feels numb, she shouts, "Yes! Wait! No! Yes! No, no! Wait, yes!" her face contorted as each of her two minds, only one of which can feel the hand, tries to answer. The researcher then hands her a sheet of paper with the words "yes" and "no" written on it and tells her to point to the correct answer. The woman stares at the sheet for a moment. Then her left forefinger stabs at "yes" and her right forefinger at "no."

To materialists such as Crick, the significance of such phenomena is obvious: when the brain is injured in certain ways, consciousness (and not necessarily perception) is impaired. Clearly, consciousness has no existence independent of what has been called the "meat machine" but is lodged firmly within it. Some audience members, however, extract different messages from Mark's video presentation, as if it were a Rorschach blot. One listener suggests that even healthy individuals experience some fracturing of the self—albeit in a much less dramatic form. A psychiatrist then wonders aloud whether the woman's two selves could be trained to get along better with each other through conflict resolution.

■ THE BINDING PROBLEM

Koch, during his presentation, strives to bring the discussion down to earth. Striding restlessly around the stage in jeans and scarlet lizard-skin cowboy boots, he delivers a high-speed, German-accented summary of the message he and Crick have promulgated over the past four years: scientists should concentrate on questions that can be experimentally resolved and leave metaphysical speculations to "late-night conversations over beer." To emphasize this point, Koch reiterates a line of the "philosopher" Clint Eastwood: "A man has gotta know his limitations."

He then details his and Crick's argument that consciousness stems from a process that combines attention with short-term memory. (Koch notes that the turn-of-the-century philosopher William James first had this insight.) The phenomenon of attention involves more than simple information processing, Koch observes. To demonstrate this point, he shows a slide of a pattern that can be viewed either as a vase or as a pair of human profiles facing each other. Although the visual input to the brain remains constant, the pattern that one is aware of, or attends to, keeps changing. What neural activity corresponds to the change in attention?

The answer to this question is complicated by the fact that "there is no single place where everything comes together" in forming a perception; even a single scene is processed by different neurons in different parts of the brain. One must therefore determine what mechanism transforms the firing of neurons scattered throughout the visual cortex into a unified perception. "This is known as the binding problem," Koch explains, noting that it is considered by many neuroscientists to be the central issue of their field.

One possible answer to the binding problem has been suggested by experiments on animals showing that neurons in different regions of the brain occasionally oscillate at the same frequency, roughly 40 times per second. Koch asks the audience to imagine the brain as a Christmas tree with billions of lights flickering apparently at random. These flickerings represent the response of the visual cortex to, say, a roomful of people. Suddenly, a subset of lights flickers at the same frequency, 40 times per second, as the mind focuses on the face of an old flame.

Koch concedes that the evidence implicating 40-hertz oscillations in awareness is tenuous; it has shown up most clearly in anesthetized cats. Another form of binding could be simple synchrony: neurons merely fire at the same time and not necessarily the same frequency. Evidence for synchrony is also trickling in from animal experiments.

After Koch's lecture, Valerie Gray Hardcastle, a philosopher at Virginia Polytechnic Institute and State University, takes the podium to proclaim that "simple solutions to the binding problem must fail." Pointing out that monitoring individual neurons may yield spurious associations, she suggests that scientists would do better to examine the behavior of populations of neurons or even of the entire brain.

Walter J. Freeman, a lean, white-bearded professor at the University of California at Berkeley, raises similar objections to Koch's remarks. Freeman's criticism is significant, because he was one of the first prominent scientists to investigate 40-hertz oscillations. He asserts that neither oscillations nor synchrony will play more than a minor role in the solution to the binding problem and that "the current wave of enthusiasm is misplaced."

Freeman advocates a more complex approach to the binding problem. Large groups of neurons, he explains, display chaotic behavior; that is, their firing seems random but actually contains a hidden order. Like all chaotic systems, these neural patterns are extremely sensitive to minute influences. The sight of a familiar face, therefore, can trigger an abrupt shift in the firing pattern corresponding to a shift in one's awareness. Freeman concedes, however, that even his theory is only—at best—one piece of the puzzle.

Freeman thinks another piece of the puzzle may be supplied by Benjamin Libet of the University of California at San Francisco, one of the few researchers who has studied consciousness by conducting extensive experiments on humans. Libet is a sharp-featured man with the narrow-eyed, somewhat defensive mien of someone who has had to fight hard for acceptance of his work. (Crick, in his new book, recalls that Libet once confided that he dared to study consciousness in human subjects only after receiving tenure.)

■ TIMING IS EVERYTHING

One set of experiments done by Libet involved patients in whose cortices electrodes had been implanted for medical reasons. Libet delivered a mild electrical pulse both to these electrodes and to the skin of the subjects. In either case, the subjects became aware of pulses only if they lasted for more than 0.5 second. When Libet stimulated the cortical neurons as much as 0.5 second before stimulating the skin, the subjects reported, paradoxically, having felt the skin pulses first.

Libet theorizes that once the volunteers became aware of the stimulation of their skin, they experienced the sensation as if they had been aware of it from the beginning and not after 0.5 second. We subjectively compensate for the time lag in our awareness of tactile sensations, Libet contends, through a process he calls "backward referral in time." He likens this ability to the one that allows an observer moving past a picket fence to maintain a constant image of the house behind it.

Libet then describes an equally surprising set of experiments on healthy subjects whose brain waves were monitored by an electroencephalograph (EEG) machine. They were instructed to flex a finger at a moment of their choosing while noting the instant of their decision as indicated on a clock. The volunteers took some 0.2 second to flex after they had decided to do so. But, according to the EEG, the subjects' brains displayed neural activity 0.3 second *before* they decided to act. In a sense, the brain made the decision to move before the mind became conscious of it.

"The actual initiation of volition may have begun even earlier in a part of the brain we weren't monitoring," Libet comments.

After Libet delivers his carefully crafted speech, one audience member asks whether his findings bear on the question of free will. "I've always been able to avoid that question," Libet replies with a grimace. He proposes that we may exert free will not by initiating intentions but by vetoing, acceding, or otherwise responding to them after they arise.

Other observers fault Libet for overgeneralizing from his data. Flanagan, the Duke philosopher who coined the term "mysterians," points out that, strictly speaking, Libet's subjects were not acting of their own free will, since they had been instructed in advance to flex their fingers. Flanagan adds that Libet's time-lag findings may hold true only for tactile responses and not for other sensory modes.

In fact, Flanagan thinks that there may be many modes of consciousness; our awareness of an odor, for example, not only stems from a different set of neurons but also is in some sense qualitatively different from our visual awareness. Flanagan contends that neuroscientists must therefore resist the temptation to look for any single mechanism—such as 40-hertz oscillations or Freeman's chaotic neural behavior or Libet's time-delay factor—that accounts for consciousness.

In his 1992 book *Consciousness Reconsidered,* Flanagan argues on behalf of a philosophy called constructive naturalism, which holds consciousness to be a common biological phenomenon occurring not only in humans but in many other animals— and certainly all the higher primates. Other adherents to his position include Daniel C. Dennett of Tufts University (author of *Consciousness Explained,* also published in 1992) and Patricia S. Churchland of the University of California at San Diego. "We say you can acquire knowledge of consciousness by triangulation," Flanagan remarks, that is, by combining neural and psychological data from experiments on humans and animals with subjective reports from humans.

■ QUANTUM MICROTUBULES

All these philosophers—and most neuroscientists—are united in their skepticism that consciousness depends in some important way on quantum effects. Since at least the 1930s some physicists have speculated that quantum mechanics and consciousness might be linked. They based their speculation on the principle that the act of measurement—which ultimately involves a conscious observer—has an effect on the outcome of quantum events. Such notions have generally involved little more than hand waving, but they have become more prominent lately because of Penrose.

Penrose has earned respect as an authority on general relativity and for inventing the geometric forms known as Penrose tiles, which fit together to form quasiperiodic patterns. In his 1989 best-seller, *The Emperor's New Mind,* he vigorously attacked the claim of artificial-intelligence proponents that computers can replicate all the attributes of humans, including consciousness.

An elfin man with a shock of black hair who manages to seem simultaneously distracted and acutely alert, Penrose summarizes the book's themes at the Tucson conference. He first tells a story about how Deep Thought, a computer that has beaten some of the world's best chess players, was stumped by an endgame situation that even a clever amateur player would have known how to handle. Penrose concludes that "what computers can't do is understand."

The key to Penrose's argument is Gödel's theorem, a 60-year-old mathematical demonstration that any moderately complex system of axioms yields statements that are self-evidently true but cannot be proved with those axioms. The implication of the theorem, according to Penrose, is that no deterministic, ruled-based system—that is, neither classical physics, computer science, nor neuroscience—can account for the mind's creative powers and ability to ascertain truth.

In fact, Penrose thinks the mind must exploit nondeterministic effects that can be described only by quantum mechanics or "a new physical theory that will bridge quantum and classical mechanics and will go beyond computation." He even suggests that nonlocality, the ability of one part of a quantum system to affect other parts instantaneously (Einstein dubbed it "spooky action at a distance"), might be the solution to the binding problem.

Although Penrose was once rather vague about where quantum effects work their magic, he now hazards a guess: microtubules, minute tunnels of protein that serve as a kind of skeleton for cells, including neurons. This endorsement delights Stuart R. Hameroff, an anesthesiologist at the University of Arizona, who organized the Tucson meeting and is the leading proponent of the microtubule hypothesis.

Hameroff, an aging hipster with a goatee and ponytail, manages to squeeze a remarkable number of scientific buzzwords into his talk on microtubules: emergent, fractal, self-organizing, dynamical. He claims to have found evidence that anesthesia eliminates consciousness by inhibiting the movement of electrons in microtubules.

Erecting a mighty theoretical edifice on this frail claim, he proposes that microtubules perform nondeterministic, quantum-based computations that "somehow" give rise to consciousness. Each neuron is thus not a simple switch but "a network within a network," Hameroff elaborates. He acknowledges that microtubules occur in most cells, not just neurons, but the implications of this fact do not faze him.

"I'm not going to contend that a paramecium is conscious, but it does show pretty intelligent behavior," he says.

Other quantum-consciousness devotees make Penrose and Hameroff look like paragons of rigor. For example, Ian N. Marshall, a British psychiatrist, presents what he believes is evidence that thought stems from quantum effects. He and several colleagues claim to have found that the ability of subjects to perform simple tests while hooked up to an EEG machine varies depending on whether that machine is plugged in or not. Their conclusion? When the EEG machine is turned on, it "observes" the brain and therefore alters its thoughts—just as observing an electron alters its course. In other words, Heisenberg's uncertainty principle applies to the entire brain.

One listener appalled by such assertions is John G. Taylor, a physicist and neural-network specialist at King's College London. He insists that all the quantum-consciousness enthusiasts, and even Penrose, ignore the most basic facts about quantum mechanics. For example, nonlocality and other quantum effects they have seized on as vital to consciousness are observed only at temperatures near absolute zero—or at any rate far below the ambient temperatures of the brain. Like most neuroscientists, Taylor also objects to the quantum approach on pragmatic grounds. Before turning to the extreme reductionist approach advocated by Penrose and others, researchers should explore possibilities that are more plausible and experimentally accessible—and that have already proved successful for explaining certain aspects of perception and memory. "If that fails, then maybe we should look elsewhere," he adds.

■ THE EXPLANATORY GAP

One of the few points of view missing at the Arizona meeting is that of Colin McGinn, a philosopher at Rutgers University and perhaps the most unflinching of all mysterians. In his 1991 book *The Problem of Consciousness,* McGinn argued that because our brains are products of evolution, they have cognitive limitations. Just as rats or monkeys cannot even conceive of quantum mechanics, so we humans may be prohibited from understanding certain aspects of existence, such as the relation between mind and matter. Consciousness, in other words, will remain forever beyond human understanding, according to McGinn.

At least one philosopher at the Arizona meeting veers dangerously close to this glum conclusion. David J. Chalmers, an Australian at Washington University who bears an uncanny resemblance to the subject of Thomas Gainsborough's painting

Blue Boy, agrees with McGinn that no strictly physical theory—whether based on quantum mechanisms or neural ones—can explain consciousness.

All physical theories, Chalmers claims, can describe only specific mental *functions*—such as memory, attention, inattention, introspection—correlating to specific physical processes in the brain. According to Chalmers, none of these theories addresses the really "hard" question posed by the existence of the mind: Why is the performance of these functions accompanied by subjective experience? After all, one can certainly imagine a world of androids that resemble humans in every respect—except that they do not have a conscious experience of the world.

Science alone cannot supply an answer to this question, Chalmers declares. Unlike McGinn, however, Chalmers holds that philosophers can and must construct a higher-level theory to bridge that "explanatory gap" between the physical and subjective realms. In fact, Chalmers has such a theory. He asserts that just as physics assumes the existence of properties of nature such as space, time, energy, charge, and mass, so must a theory of consciousness posit the existence of a new fundamental property: information. The concept of information, Chalmers explains, has aspects that are both physical and "phenomenal" (a philosopher's term that is roughly equivalent to "experiential" or to "subjective").

Koch finds such arguments irksome. He notes that if everyone shared the belief of McGinn and Chalmers that science cannot solve consciousness, the prophecy would be self-fulfilling. Science may not be able to resolve all the mysteries of the mind, Koch concedes, but philosophy has a much slimmer chance of providing lasting insights about the mind-body problem or the question of free will. He adds that when considering such ancient conundrums, philosophers should heed the advice offered by their illustrious forebear Ludwig Wittgenstein: "Whereof one cannot speak, thereof one must be silent."

Yet of all the outcomes of the surging interest in consciousness, the least likely is silence. The evening after Chalmers gives his speech in Tucson (which is extremely well received), Koch confronts him to complain that his "double-aspect theory of information" is untestable and therefore useless. "Why don't you just say that when you have a brain the Holy Ghost comes down and makes you conscious!" Koch exclaims. Such a theory is unnecessarily complicated, Chalmers responds drily, and it would not accord with his own subjective experience. "But how do I know your subjective experience is the same as mine?" Koch spurts. "How do I even know you're conscious?"

Later, Koch and Chalmers head off to the hotel bar to continue their discussion over beer. They come to a reconciliation of sorts. Koch expresses interest in Chalmers's views on computation and cognition—a paper on which Chalmers just happens to have in his knapsack. Chalmers concedes that perhaps neuroscience may still provide direction and inspiration for philosophy.

The field of consciousness studies is struggling with its own binding problem. Yet from encounters such as this, progress may emerge—just as the interactions between neurons rather than their individual properties give rise to the miracle of the mind.

—July 1994

THE PROBLEM

OF CONSCIOUSNESS

It can now be approached by scientific investigation of the
visual system. The solution will require a close collaboration
among psychologists, neuroscientists, and theorists

Francis Crick and Christof Koch

The overwhelming question in neurobiology today is the relation between the mind and the brain. Everyone agrees that what we know as mind is closely related to certain aspects of the behavior of the brain, not to the heart, as Aristotle thought. Its most mysterious aspect is consciousness or awareness, which can take many forms, from the experience of pain to self-consciousness. In the past the mind (or soul) was often regarded, as it was by Descartes, as something immaterial, separate from the brain but interacting with it in some way. A few neuroscientists, such as Sir John Eccles, still assert that the soul is distinct from the body. But most neuroscientists now believe that all aspects of mind, including its most puzzling attribute—consciousness or awareness—are likely to be explainable in a more materialistic way as the behavior of large sets of interacting neurons. As William James, the father of American psychology, said a century ago, consciousness is not a thing but a process.

Exactly what the process is, however, has yet to be discovered. For many years after James penned *The Principles of Psychology,* consciousness was a taboo concept in American psychology because of the dominance of the behaviorist movement. With the advent of cognitive science in the mid-1950s, it became possible once more for psychologists to consider mental processes as opposed to merely observing behavior. In spite of these changes, until recently most cognitive scientists ignored consciousness, as did almost all neuroscientists. The problem was felt to be either purely "philosophical" or too elusive to study experimentally. It would not have been easy for a neuroscientist to get a grant just to study consciousness.

In our opinion, such timidity is ridiculous, so a few years ago we began to think about how best to attack the problem scientifically. How to explain mental events as being caused by the firing of large sets of neurons? Although there are those who believe such an approach is hopeless, we feel it is not productive to worry too much over aspects of the problem that cannot be solved scientifically or, more precisely, cannot be solved solely by using existing scientific ideas. Radically new concepts may indeed be needed—recall the modifications of scientific thinking forced on us by quantum mechanics. The only sensible approach is to press the experimental attack until we are confronted with dilemmas that call for new ways of thinking.

There are many possible approaches to the problem of consciousness. Some psychologists feel that any satisfactory theory should try to explain as many aspects of consciousness as possible, including emotion, imagination, dreams, mystical experiences, and so on. Although such an all-embracing theory will be necessary in the long run, we thought it wiser to begin with the particular aspect of consciousness that is likely to yield most easily. What this aspect may be is a matter of personal judgment. We selected the mammalian visual system because humans are very visual animals and because so much experimental and theoretical work has already been done on it.

It is not easy to grasp exactly what we need to explain, and it will take many careful experiments before visual consciousness can be described scientifically. We did not attempt to define consciousness itself because of the dangers of premature definition. (If this seems like a copout, try defining the word "gene"—you will not find it easy.) Yet the experimental evidence that already exists provides enough of a glimpse of the nature of visual consciousness to guide research. In this article, we will attempt to show how this evidence opens the way to attack this profound and intriguing problem.

Visual theorists agree that the problem of visual consciousness is ill posed. The mathematical term "ill posed" means that additional constraints are needed to solve the problem. Although the main function of the visual system is to perceive objects and events in the world around us, the information available to our eyes is not sufficient by itself to provide the brain with its unique interpretation of the visual world. The brain must use past experience (either its own or that of our distant ancestors, which is embedded in our genes) to help interpret the information coming into our eyes. An example would be the derivation of the three-dimensional representation of the world from the two-dimensional signals falling onto the retinas of our two eyes or even onto one of them.

Visual theorists also would agree that seeing is a constructive process, one in which the brain has to carry out complex activities (sometimes called computations) in order to decide which interpretation to adopt of the ambiguous visual input. "Computation" implies that the brain acts to form a symbolic representation of the visual world, with a mapping (in the mathematical sense) of certain aspects of that world onto elements in the brain.

Ray Jackendoff of Brandeis University postulates, as do most cognitive scientists, that the computations carried out by the brain are largely unconscious and that what we become aware of is the result of these computations. But while the customary view is that this awareness occurs at the highest levels of the computational system, Jackendoff has proposed an intermediate-level theory of consciousness.

What we see, Jackendoff suggests, relates to a representation of surfaces that are directly visible to us, together with their outline, orientation, color, texture, and movement. (This idea has similarities to what the late David C. Marr of the Massachusetts Institute of Technology called a "2½-dimensional sketch." It is more than a two-dimensional sketch because it conveys the orientation of the visible surfaces. It is less than three-dimensional because depth information is not explicitly represented.) In the next stage this sketch is processed by the brain to produce a three-dimensional representation. Jackendoff argues that we are not visually aware of this three-dimensional representation.

An example may make this process clearer. If you look at a person whose back is turned to you, you can see the back of the head but not the face. Nevertheless, your brain infers that the person has a face. We can deduce as much because if that person turned around and had no face, you would be very surprised.

The viewer-centered representation that corresponds to the visible back of the head is what you are vividly aware of. What your brain infers about the front would come from some kind of three-dimensional representation. This does not mean that information flows only from the surface representation to the three-dimensional one; it almost certainly flows in both directions. When you imagine the front of the face, what you are aware of is a surface representation generated by information from the three-dimensional model.

It is important to distinguish between an explicit and an implicit representation. An explicit representation is something that is symbolized without further processing. An implicit representation contains the same information but requires further processing to make it explicit. The pattern of colored dots on a television screen, for example, contains an implicit representation of objects (say, a person's face), but only the dots and their locations are explicit. When you see a face on the screen, there must be neurons in your brain whose firing, in some sense, symbolizes that face.

We call this pattern of firing neurons an active representation. A latent representation of a face must also be stored in the brain, probably as a special pattern of synaptic connections between neurons. For example, you probably have a representation of the Statue of Liberty in your brain, a representation that usually is inactive. If you do think about the Statue, the representation becomes active, with the relevant neurons firing away.

An object, incidentally, may be represented in more than one way—as a visual image, as a set of words and their related sounds, or even as a touch or a smell. These different representations are likely to interact with one another. The representation is likely to be distributed over many neurons, both locally and more globally. Such a representation may not be as simple and straightforward as uncritical introspection might indicate. There is suggestive evidence, partly from studying how neurons fire in various parts of a monkey's brain and partly from examining the effects of certain types of brain damage in humans, that different aspects of a face—and of the implications of a face—may be represented in different parts of the brain.

First, there is the representation of a face as a face: two eyes, a nose, a mouth, and so on. The neurons involved are usually not too fussy about the exact size or position of this face in the visual field, nor are they very sensitive to small changes in its orientation. In monkeys, there are neurons that respond best when the face is turning in a particular direction, while others seem to be more concerned with the direction in which the eyes are gazing.

Then there are representations of the parts of a face, as separate from those for the face as a whole. Further, the implications of seeing a face, such as that person's sex, the facial expression, the familiarity or unfamiliarity of the face, and in particular whose face it is, may each be correlated with neurons firing in other places.

What we are aware of at any moment, in one sense or another, is not a simple matter. We have suggested that there may be a very transient form of fleeting awareness that represents only rather simple features and does not require an attentional mechanism. From this brief awareness the brain constructs a viewer-centered representation—what we see vividly and clearly—that does require attention. This in turn probably leads to three-dimensional object representations and thence to more cognitive ones.

Representations corresponding to vivid consciousness are likely to have special properties. William James thought that consciousness involved both attention and short-term memory. Most psychologists today would agree with this view. Jackendoff writes that consciousness is "enriched" by attention, implying that whereas attention may not be essential for certain limited types of consciousness, it is necessary for full consciousness. Yet it is not clear exactly which forms of memory are involved. Is long-term memory needed? Some forms of acquired knowledge are so embedded in the machinery of neural processing that they are almost certainly used in becoming aware of something. On the other hand, there is evidence from studies of brain-damaged patients that the ability to lay down new long-term episodic memories is not essential for consciousness to be experienced.

It is difficult to imagine that anyone could be conscious if he or she had no memory whatsoever of what had just happened, even an extremely short one. Visual psychologists talk of iconic memory, which lasts for a fraction of a second, and working memory (such as that used to remember a new telephone number) that lasts for only a few seconds unless it is rehearsed. It is not clear whether both of these are essential for consciousness. In any case, the division of short-term memory into these two categories may be too crude.

If these complex processes of visual awareness are localized in parts of the brain, which processes are likely to be where? Many regions of the brain may be involved, but it is almost certain that the cerebral neocortex plays a dominant role. Visual information from the retina reaches the neocortex mainly by way of a part of the thalamus (the lateral geniculate nucleus); another significant visual pathway from the retina is to the superior colliculus, at the top of the brain stem.

The cortex in humans consists of two intricately folded sheets of nerve tissue, one on each side of the head. These sheets are connected by a large tract of about half a billion axons called the corpus callosum. It is well known that if the corpus callosum is cut, as is done for certain cases of intractable epilepsy, one side of the brain is not aware of what the other side is seeing. In particular, the left side of the brain (in a right-handed person) appears not to be aware of visual information received exclusively by the right side. This shows that none of the information required for visual awareness can reach the other side of the brain by traveling down to the brain stem and, from there, back up. In a normal person, such information can get to the other side only by using the axons in the corpus callosum.

A different part of the brain—the hippocampal system—is involved in one-shot, or episodic, memories that, over weeks and months, it passes on to the neocortex. This system is so placed that it receives inputs from, and projects to, many parts of the brain. Thus, one might suspect that the hippocampal system is the essential seat of consciousness. This is not the case: evidence from studies of patients with damaged brains shows that this system is not essential for visual awareness, although naturally a patient lacking one is severely handicapped in everyday life because he cannot remember anything that took place more than a minute or so in the past.

In broad terms, the neocortex of alert animals probably acts in two ways. By building on crude and somewhat redundant wiring, produced by our genes and by embryonic processes, the neocortex draws on visual and other experience to slowly "rewire" itself to create categories (or "features") it can respond to. A new category is not fully created in the neocortex after exposure to only one example of it, although some small modifications of the neural connections may be made.

The second function of the neocortex (at least of the visual part of it) is to respond extremely rapidly to incoming signals. To do so, it uses the categories it has learned and tries to find the combinations of active neurons that, on the basis of its past experience, are most likely to represent the relevant objects and events in the visual world at that moment. The formation of such coalitions of active neurons may also be influenced by biases coming from other parts of the brain: for example, signals telling it what best to attend to or high-level expectations about the nature of the stimulus.

Consciousness, as James noted, is always changing. These rapidly formed coalitions occur at different levels and interact to form even broader coalitions. They are transient, lasting usually for only a fraction of a second. Because coalitions in the visual system are the basis of what we see, evolution has seen to it that they form as fast as possible; otherwise, no animal could survive. The brain is handicapped in forming neuronal coalitions rapidly because, by computer standards, neurons act very slowly. The brain compensates for this relative slowness partly by using very many neurons, simultaneously and in parallel, and partly by arranging the system in a roughly hierarchical manner.

If visual awareness at any moment corresponds to sets of neurons firing, then the obvious question is: Where are these neurons located in the brain, and in what way are they firing? Visual awareness is highly unlikely to occupy all the neurons in the neocortex that are firing above their background rate at a particular moment. We would expect that, theoretically, at least some of these neurons would be involved in doing computations—trying to arrive at the best coalitions—whereas others would express the results of these computations, in other words, what we see.

Fortunately, some experimental evidence can be found to back up this theoretical conclusion. A phenomenon called binocular rivalry may help identify the neurons whose firing symbolizes awareness. This phenomenon can be seen in dramatic form in an exhibit prepared by Sally Duensing and Bob Miller at the Exploratorium in San Francisco.

■ CONFLICTING INPUTS

Binocular rivalry occurs when each eye has a different visual input relating to the same part of the visual field. The early visual system of the left side of the brain receives an input from both eyes but sees only the part of the visual field to the right of the fixation point. The converse is true for the right side. If these two conflicting inputs are rivalrous, one sees not the two inputs superimposed but first one input, then the other, and so on in alternation.

In the exhibit, called "The Cheshire Cat," viewers put their heads in a fixed place and are told to keep the gaze fixed. By means of a suitably placed mirror, one of the eyes can look at another person's face, directly in front, while the other eye sees a blank white screen to the side. If the viewer waves a hand in front of this plain screen at the same location in his or her visual field occupied by the face, the face is wiped out. The movement of the hand, being visually very salient, has captured the brain's attention. Without attention the face cannot be seen. If the viewer moves the eyes, the face reappears.

In some cases, only part of the face disappears. Sometimes, for example, one eye, or both eyes, will remain. If the viewer looks at the smile on the person's face, the face may disappear, leaving only the smile. For this reason the effect has been called the Cheshire Cat effect, after the cat in Lewis Carroll's *Alice's Adventures in Wonderland*.

Although it is very difficult to record activity in individual neurons in a human brain, such studies can be done in monkeys. A simple example of binocular rivalry has been studied in a monkey by Nikos K. Logothetis and Jeffrey D. Schall, both then at M.I.T. They trained a macaque to keep its eyes still and to signal whether it is seeing upward or downward movement of a horizontal grating. To produce rivalry, upward movement is projected into one of the monkey's eyes and downward movement into the other, so that the two images overlap in the visual field. The monkey signals that it sees up and down movements alternatively, just as humans would. Even though the motion stimulus coming into the monkey's eyes is always the same, the monkey's percept changes every second or so.

Cortical area MT (which some researchers prefer to label V5) is an area mainly concerned with movement. What do the neurons in MT do when the monkey's percept is sometimes up and sometimes down? (The researchers studied only the monkey's first response.) The simplified answer—the actual data are rather more messy—is that whereas the firing of some of the neurons correlates with the changes in the percept, for others the average firing rate is relatively unchanged and independent of which direction of movement the monkey is seeing at that moment. Thus, it is unlikely that the firing of all the neurons in the visual neocortex at one particular moment corresponds to the monkey's visual awareness. Exactly which neurons do correspond to awareness remains to be discovered.

We have postulated that when we clearly see something, there must be neurons actively firing that stand for what we see. This might be called the activity principle. Here, too, there is some experimental evidence. One example is the firing of neurons in a specific cortical visual area in response to illusory contours. Another and perhaps more striking case is the filling in of the blind spot. The blind spot in each eye is caused by the lack of photoreceptors in the area of the retina where the optic nerve leaves the retina and projects to the brain. Its location is about 15 degrees from the fovea (the visual center of the eye). Yet if you close one eye, you do not see a hole in your visual field.

Philosopher Daniel C. Dennett of Tufts University is unusual among philosophers in that he is interested both in psychology and in the brain. This interest is much to be welcomed. In a recent book, *Consciousness Explained,* he has argued that it is wrong to talk about filling in. He concludes, correctly, that "an absence of information is not the same as information about an absence." From this general principle he argues that the brain does not fill in the blind spot but rather ignores it.

Dennett's argument by itself, however, does not establish that filling in does not occur; it only suggests that it might not. Dennett also states that "your brain has no machinery for [filling in] at this location." This statement is incorrect. The primary visual cortex lacks a direct input from one eye, but normal "machinery" is there to deal with the input from the other eye. Ricardo Gattas and his colleagues at the Federal University of Rio de Janeiro have shown that in the macaque some of the neurons in the blind-spot area of the primary visual cortex do respond to input from both eyes, probably assisted by inputs from other parts of the cortex. Moreover, in the case of simple filling in, some of the neurons in that region respond as if they are actively filling in [*see color plate 4*].

Thus, Dennett's claim about blind spots is incorrect. In addition, psychological experiments by Vilayanur S. Ramachandran have shown that what is filled in can be quite complex depending on the overall context of the visual scene. How, he argues, can your brain be ignoring something that is in fact commanding attention?

Filling in, therefore, is not to be dismissed as nonexistent or unusual. It probably represents a basic interpolation process that can occur at many levels in the neocortex. It is, incidentally, a good example of what is meant by a constructive process.

How can we discover the neurons whose firing symbolizes a particular percept? William T. Newsome and his colleagues at Stanford University have done a series of brilliant experiments on neurons in cortical area MT of the macaque's brain. By studying a neuron in area MT, we may discover that it responds best to very specific visual features having to do with motion. A neuron, for instance, might fire strongly in response to the movement of a bar in a particular place in the visual field, but only when the bar is oriented at a certain angle, moving in one of the two directions perpendicular to its length within a certain range of speed.

It is technically difficult to excite just a single neuron, but it is known that neurons that respond to roughly the same position, orientation, and direction of movement of a bar tend to be located near one another in the cortical sheet. The experimenters taught the monkey a simple task in movement discrimination using a mixture of dots, some moving randomly, the rest all in one direction. They showed that electrical stimulation of a small region in the right place in cortical area MT would bias the monkey's motion discrimination, almost always in the expected direction.

Thus, the stimulation of these neurons can influence the monkey's behavior and probably its visual percept. Such experiments do not, however, show decisively that the firing of such neurons is the exact neural correlate of the percept. The correlate could be only a subset of the neurons being activated. Or perhaps the real correlate is the firing of neurons in another part of the visual hierarchy that are strongly influenced by the neurons activated in area MT.

These same reservations apply also to cases of binocular rivalry. Clearly, the problem of finding the neurons whose firing symbolizes a particular percept is not going to be easy. It will take many careful experiments to track them down even for one kind of percept.

It seems obvious that the purpose of vivid visual awareness is to feed into the cortical areas concerned with the implications of what we see; from there the information shuttles on the one hand to the hippocampal system, to be encoded (temporarily) into long-term episodic memory, and on the other to the planning levels of the motor system. But is it possible to go from a visual input to a behavioral output without any relevant visual awareness?

That such a process can happen is demonstrated by the remarkable class of patients with "blindsight." These patients, all of whom have suffered damage to their visual cortex, can point with fair accuracy at visual targets or track them with their eyes while vigorously denying seeing anything. In fact, the patients are as surprised as their doctors by their abilities. The amount of information that "gets through," however, is limited: blindsight patients have some ability to respond to wavelength, orientation, and motion, yet they cannot distinguish a triangle from a square.

It is naturally of great interest to know which neural pathways are being used in these patients. Investigators originally suspected that the pathway ran through the superior colliculus. Recent experiments suggest that a direct albeit weak connection may be involved between the lateral geniculate nucleus and other visual areas in the cortex. It is unclear whether an intact primary visual cortex region is essential for immediate visual awareness. Conceivably the visual signal in blindsight is so weak that the neural activity cannot produce awareness, although it remains strong enough to get through to the motor system.

Normal-seeing people regularly respond to visual signals without being fully aware of them. In automatic actions, such as swimming or driving a car, complex but stereotypical actions occur with little, if any, associated visual awareness. In other cases, the information conveyed is either very limited or very attenuated. Thus, while we can function without visual awareness, our behavior without it is rather restricted.

Clearly, it takes a certain amount of time to experience a conscious percept. It is difficult to determine just how much time is needed for an episode of visual awareness, but one aspect of the problem that can be demonstrated experimentally is that signals received close together in time are treated by the brain as simultaneous.

A disk of red light is flashed for, say, 20 milliseconds, followed immediately by a 20-millisecond flash of green light in the same place. The subject reports that he did not see a red light followed by a green light. Instead he saw a yellow light, just as he would have if the red and the green light had been flashed simultaneously. Yet the subject could not have experienced yellow until after the information from the green flash had been processed and integrated with the preceding red one.

Experiments of this type led psychologist Robert Efron, now at the University of California at Davis, to conclude that the processing period for perception is about 60 to 70 milliseconds. Similar periods are found in experiments with tones in the auditory system. It is always possible, however, that the processing times may be different in higher parts of the visual hierarchy and in other parts of the brain. Processing is also more rapid in trained, compared with naive, observers.

Because it appears to be involved in some forms of visual awareness, it would help if we could discover the neural basis of attention. Eye movement is a form of attention, since the area of the visual field in which we see with high resolution is remarkably small, roughly the area of the thumbnail at arm's length. Thus, we move our eyes to gaze directly at an object in order to see it more clearly. Our eyes usually move three or four times a second. Psychologists have shown, however, that there appears to be a faster form of attention that moves around, in some sense, when our eyes are stationary.

The exact psychological nature of this faster attentional mechanism is at present controversial. Several neuroscientists, however, including Robert Desimone and his colleagues at the National Institute of Mental Health, have shown that the rate of firing of certain neurons in the macaque's visual system depends on what the monkey is attending to in the visual field. Thus, attention is not solely a psychological concept; it also has neural correlates that can be observed. A number of researchers have found that the pulvinar, a region of the thalamus, appears to be involved in visual attention. We would like to believe that the thalamus deserves to be called "the organ of attention," but this status has yet to be established.

■ ATTENTION AND AWARENESS

The major problem is to find what activity in the brain corresponds directly to visual awareness. It has been speculated that each cortical area produces awareness of only those visual features that are "columnar," or arranged in the stack or column of neurons perpendicular to the cortical surface. Thus, the primary visual cortex could code for orientation and area MT for motion. So far experimentalists have not found one particular region in the brain where all the information needed for visual awareness appears to come together. Dennett has dubbed such a hypothetical place "The Cartesian Theater." He argues on theoretical grounds that it does not exist.

Awareness seems to be distributed not just on a local scale, but more widely over the neocortex. Vivid visual awareness is unlikely to be distributed over every cortical area because some areas show no response to visual signals. Awareness might, for example, be associated with only those areas that connect back directly to the primary visual cortex or alternatively with those areas that project into one another's layer 4. (The latter areas are always at the same level in the visual hierarchy.)

The key issue, then, is how the brain forms its global representations from visual signals. If attention is indeed crucial for visual awareness, the brain could form representations by attending to just one object at a time, rapidly moving from one object to the next. For example, the neurons representing all the different aspects of the attended object could all fire together very rapidly for a short period, possibly in rapid bursts.

This fast, simultaneous firing might not only excite those neurons that symbolized the implications of that object but also temporarily strengthen the relevant synapses so that this particular pattern of firing could be quickly recalled—a form of short-term memory. If only one representation needs to be held in short-term memory, as in remembering a single task, the neurons involved may continue to fire for a period.

A problem arises if it is necessary to be aware of more than one object at exactly the same time. If all the attributes of two or more objects were represented by neurons firing rapidly, their attributes might be confused. The color of one might become attached to the shape of another. This happens sometimes in very brief presentations.

Some time ago Christoph von der Malsburg, now at the Ruhr-Universität Bochum, suggested that this difficulty would be circumvented if the neurons associated with any one object all fired in synchrony (that is, if their times of firing were correlated) but out of synchrony with those representing other objects. Recently two groups in Germany reported that there does appear to be correlated firing between neurons in the visual cortex of the cat, often in a rhythmic manner, with a frequency in the 35- to 75-hertz range, sometimes called 40-hertz, or g, oscillation.

Von der Malsburg's proposal prompted us to suggest that this rhythmic and synchronized firing might be the neural correlate of awareness and that it might serve to bind together activity concerning the same object in different cortical areas. The matter is still undecided, but at present the fragmentary experimental evidence does rather little to support such an idea. Another possibility is that the 40-hertz oscillations may help distinguish figure from ground or assist the mechanism of attention.

■ CORRELATES OF CONSCIOUSNESS

Are there some particular types of neurons, distributed over the visual neocortex, whose firing directly symbolizes the content of visual awareness? One very simplistic hypothesis is that the activities in the upper layers of the cortex are largely unconscious ones, whereas the activities in the lower layers (layers 5 and 6) mostly correlate with consciousness. We have wondered whether the pyramidal neurons in layer 5 of the neocortex, especially the larger ones, might play this latter role.

These are the only cortical neurons that project right out of the cortical system (that is, not to the neocortex, the thalamus, or the claustrum). If visual awareness represents the results of neural computations in the cortex, one might expect that what the cortex sends elsewhere would symbolize those results. Moreover, the neurons in layer 5 show a rather unusual propensity to fire in bursts. The idea that layer 5 neurons may directly symbolize visual awareness is attractive, but it still is too early to tell whether there is anything in it.

Visual awareness is clearly a difficult problem. More work is needed on the psychological and neural basis of both attention and very short term memory. Studying the neurons when a percept changes, even though the visual input is constant, should be a powerful experimental paradigm. We need to construct neurobiological theories of visual awareness and test them using a combination of molecular, neurobiological, and clinical imaging studies.

We believe that once we have mastered the secret of this simple form of awareness, we may be close to understanding a central mystery of human life: how the physical events occurring in our brains while we think and act in the world relate to our subjective sensations—that is, how the brain relates to the mind.

—September 1992

Postscript: There have been several relevant developments since this article was first published. It now seems likely that there are rapid "on-line" systems for stereotyped motor responses such as hand or eye movement. These systems are unconscious and lack memory. Conscious seeing, on the other hand, seems to be slower and more subject to visual illusions. The brain needs to form a conscious representation of the visual scene that it then can use for many different actions or thoughts. Exactly how all these pathways work and how they interact is far from clear.

There have been more experiments on the behavior of neurons that respond to bistable visual percepts, such as binocular rivalry, but it is probably too early to draw firm conclusions from them about the exact neural correlates of visual consciousness. We have suggested on theoretical grounds based on the neuroanatomy of the macaque monkey that primates are not directly aware of what is happening in the primary visual cortex, even though most of the visual information flows through it. This hypothesis is supported by some experimental evidence, but it is still controversial.

—1997

ABOUT THE AUTHORS[*]

Russell A. Barkley is director of psychology and professor of psychiatry and neurology at the University of Massachusetts Medical Center in Worcester. He received his B.A. from the University of North Carolina at Chapel Hill and his M.S. and Ph.D. from Bowling Green State University. He has studied ADHD for nearly twenty-five years.

William Byne is a research associate at the Albert Einstein College of Medicine of Yeshiva University in New York City, where he investigates the brain structure of humans and other primates, as well as an attending psychiatrist at the New York Psychiatric Institute. He received his Ph.D. in 1985 from the Neurosciences Training Program at the University of Wisconsin at Madison, and his M.D. in 1989 from Einstein. He is also a psychiatrist in private practice.

David J. Chalmers studied mathematics at Adelaide University and as a Rhodes Scholar at the University of Oxford, but a fascination with consciousness led him into philosophy and cognitive science. He has a Ph.D. in these fields from Indiana University and is currently in the department of philosophy at the University of California at Santa Cruz.

Francis Crick is the co-discoverer, with James Watson, of the double helical structure of DNA. While at the Medical Research Council Laboratory of Molecular Biology in Cambridge, he worked on the genetic code and on developmental biology. Since 1976, he has been at the Salk Institute for Biological Studies in San Diego. His main interest lies in understanding the visual system of mammals.

Antonio R. Damasio is professor and head of the department of neurology at the University of Iowa College of Medicine and adjunct professor at the Salk Institute for Biological Studies in San Diego. He received his M.D. and doctorate from the University of Lisbon.

Hanna Damasio holds an M.D. from the University of Lisbon. She is professor of neurology and director of the Laboratory for Neuroimaging and Human Neuroanatomy at the University of Iowa.

John C. DeFries directs the University of Colorado's Institute for Behavioral Genetics and the university's Colorado Learning Disabilities Research Center. He has collaborated with Robert Plomin for more than twenty years, beginning in 1974 when Plomin, too, worked at the University of Colorado.

Uta Frith is a senior scientist in the Cognitive Development Unit of the Medical Research Council in London. Born in Germany, she took a degree in psychology in 1964 at the University of the Saarland in Saarbrücken, where she also studied art history. Four years later she obtained her Ph.D. in psychology at the University of London.

Michael S. Gazzaniga is professor of cognitive neuroscience and director of the Center for Cognitive Neuroscience at Dartmouth College. He received his Ph.D. at the California Institute of Technology, where he, Roger W. Sperry, and Joseph E. Bogen initiated split-brain studies. Gazzaniga has published in many areas and is credited with launching the field of cognitive neuroscience in the early 1980s.

W. Wayt Gibbs is a former writer for *Scientific American*.

Patricia S. Goldman-Rakic received a Ph.D. from the University of California at Los Angeles in 1963. Two years later she joined the Intramural Research Program of the National Institute of Mental Health. In 1979 she moved to the Yale University School of Medicine, where she is a professor of neuroscience. Goldman-Rakic sits on several national advisory boards and is a member of the National Academy of Sciences.

Linda S. Gottfredson is professor of educational studies at the University of Delaware, where she has been since 1986, and co-directs the Delaware–Johns Hopkins Project for Study of Intelligence and Society. She trained as a sociologist, and her earliest work focused on career development. In the mid-1980s, while at Johns Hopkins University, she published several influential articles describing how intelligence shapes vocational choice and self-perception. Gottfredson also organized the 1994 treatise "Mainstream Science on Intelligence," an editorial with more than fifty signatories that first appeared in the *Wall Street Journal* in response to the controversy surrounding the publication of *The Bell Curve*.

Dean H. Hamer received his Ph.D. in biological chemistry from Harvard in 1977. He is now at the National Institutes of Health, where he is chief of the section on gene structure and regulation at the National Cancer Institute. He studies the role of genes both in sexual orientation and in complex medical conditions, including progression of HIV and Kaposi's sarcoma.

Robert D. Hawkins received a B.A. from Stanford University and a Ph.D. in experimental psychology from the University of California at San Diego. He is associate professor at the Center for Neurobiology and Behavior at Columbia University. He has collaborated on studies of the neurobiology of learning with Eric R. Kandel.

John Horgan is a former senior writer at *Scientific American*. He is a freelance writer for *New Republic*, *Discover*, *The New York Times Book Review*, *Omni*, *Science*, and *New Scientist*.

Kay Redfield Jamison is professor of psychiatry at the Johns Hopkins University School of Medicine. She is the author of *Touched with Fire: Manic-Depressive Illness and the Artistic Temperament*. Jamison is a member of the National Advisory Council for Human Genome Research and clinical director of the Dana Consortium on the Genetic Basis of Manic-Depressive Illness. She has also written and produced a series of public television specials about manic-depressive illness and the arts.

Ned H. Kalin, a clinician and researcher, is professor of psychiatry and psychology and chairman of the department of psychiatry at the University of Wisconsin Medical School. He is also a scientist at the Wisconsin Regional Primate Research Center and the Harlow Primate Laboratory at the university. He earned his B.S. degree in 1972 from Pennsylvania State University and his M.D. in 1976 from Jefferson Medical College in Philadelphia. Before joining his current departments, he completed a residency program in psychiatry at Wisconsin and a postdoctoral fellowship in clinical neuropharmacology at the National Institute of Mental Health.

Eric R. Kandel is University Professor at the College of Physicians and Surgeons of Columbia University and senior investigator at the Howard Hughes Medical Institute. He received an A.B. from Harvard College, an M.D. from the New York University School of Medicine, and psychiatric training at Harvard Medical School. He has collaborated on studies of the neurobiology of learning with Robert D. Hawkins.

Doreen Kimura is professor of psychology and honorary lecturer in the department of clinical neurological sciences at the University of Western Ontario in London. Kimura, a fellow of the Royal Society of Canada, received the 1992 John

Dewan Award for outstanding research from the Ontario Mental Health Foundation.

Christof Koch was awarded his Ph.D. in biophysics by the University of Tübingen. After spending four years at the Massachusetts Institute of Technology he joined the California Institute of Technology where he is now professor of computation and neural systems. He is currently studying how single brain cells process information and the neural basis of motion perception, visual attention, and awareness.

Joseph E. LeDoux, who is a professor of neural science and psychology at New York University, is the recipient of two National Institute of Mental Health distinctions: a Merit Award and a Research Scientist Development Award. He has also received an Established Investigator Award from the American Heart Association.

Simon LeVay earned a doctorate in neuroanatomy at the University of Göttingen in Germany. In 1971 he went to Harvard University to work with David Hubel and Torsten Wiesel on the brain's visual system. He moved to the Salk Institute for Biological Studies in San Diego in 1984 to head the vision laboratory. In 1992 he left Salk to found the Institute of Gay and Lesbian Education.

Elizabeth F. Loftus is professor of psychology and adjunct professor of law at the University of Washington. She received her Ph.D. in psychology from Stanford University in 1970. Her research has focused on human memory, eyewitness testimony, and courtroom procedure. She has published eighteen books and has served as expert witness or consultant in hundreds of trials. Her book *Eyewitness Testimony* won a National Media Award from the American Psychological Foundation. She has received honorary doctorates from Miami University, Leiden University, and John Jay College of Criminal Justice.

Charles B. Nemeroff is Reunette W. Harris Professor and chairman of the department of psychiatry and behavioral sciences at the Emory University School of Medicine. He earned his M.D. and Ph.D. (in neurobiology) from the University of North Carolina at Chapel Hill and received psychiatry training there and at Duke University, where he joined the faculty. In 1991 he moved to Emory. Nemeroff has won several awards for his research in biological psychiatry and is immediate past president of the American College of Neuropsychopharmacology.

Robert Plomin is research professor of behavioral genetics and deputy director of the Social, Genetic, and Developmental Psychiatry Research Center at the Institute of Psychiatry in London. He has collaborated for more than twenty years with John C. DeFriess, beginning at the University of Colorado at Boulder in 1974.

Marcus E. Raichle is professor of neurology, radiology, and neurobiology as well as a senior fellow of the McDonnell Center for Studies of Higher Brain Function at the Washington University School of Medicine in St. Louis. He received his B.S. and M.D. degrees from the University of Washington in Seattle. He began researching brain metabolism and circulation when he was a neurology resident at the New York Hospital–Cornell Medical Center.

Peter Riederer heads the Laboratory of Clinical Neurochemistry and is professor of clinical neurochemistry at the University of Wurzburg in Germany. He has collaborated with Moussa B. H. Youdim since 1974 and shares with him the Claudius Galenus Gold Medal for the development of the anti-Parkinson's drug selegiline.

Dennis J. Selkoe is co-director of the Center for Neurological Diseases at Brigham and Women's Hospital in Boston and professor of neurology and neuroscience at Harvard Medical School. He received his bachelor's degree from Columbia University and his medical degree from the University of Virginia. After research training at the National Institutes of Health, Selkoe completed a fellowship in biochemistry and neuronal cell biology at Harvard Medical School and Children's Hospital in Boston. In 1978 he founded an independent research laboratory for the study of Alzheimer's disease.

Carla J. Shatz is professor of neurobiology at the University of California at Berkeley, a position she took after many years at Stanford University. She graduated from Radcliffe College and received a master's degree in physiology from University College, London, and a Ph.D. in neurobiology from Harvard Medical School. Her studies of the development of connections in the mammalian visual system have gained her many honors, including her election to the American Academy of Arts and Sciences.

Ellen Winner received her Ph.D. in psychology from Harvard University in 1978 and is currently professor of psychology at Boston College as well as senior research associate with Harvard's Project Zero, which researches the psychological aspects of the arts. She is the author of several books, including *Gifted Children: Myths and Realities*.

Moussa B. H. Youdim, a pioneer in the development of monoamine oxidase inhibitors for the treatment of Parkinson's disease and depression, is professor of pharmacology at Technion—Israel Institute of Technology in Haifa, Israel. He is also the director of the Eve Topf and U.S. National Parkinson's Disease Foundation's Centers of Excellence for Neurodegenerative Diseases, both at Technion, and a Fogarty Scholar in Residence at the U.S. National Institutes of Health, where he spends three months every year. He has collaborated with Peter Riederer since 1974 and shares with him the Claudius Galenus Gold Medal for the development of the anti-Parkinson's drug selegiline.

Semir Zeki is professor of neurobiology at the University of London. He obtained his doctorate from University College, London, and did his postdoctoral work at the National Institute of Mental Health in Washington, D.C., and at the University of Wisconsin at Madison. Zeki has also served as a visiting professor at several American and European universities.

*This information was compiled at the time the articles were originally published in *Scientific American;* some biographies may not be completely up-to-date.

INDEX